FACILITIES PLANNING

FACILITIES PLANNING

The User Requirements Method

Roger L. Brauer, Ph.D.

amacom

American Management Association

*This book is available at a special
discount when ordered in bulk quantities.
For information, contact Special Sales Department,
AMACOM, a division of American Management Association,
135 West 50th Street, New York, NY 10020.*

Library of Congress Cataloging-in-Publication Data

*Brauer, Roger L.
Facilities planning.*

*Bibliography: p.
Includes index.
1. Facility management. 2. Factories—Design and construction. I. Title.
TS177.B7 1986 725'.4 85-47679
ISBN 0-8144-5855-6*

Printing number

10 9 8 7 6 5 4 3 2 1

To
Charlotte, Michelle, and David

Preface

The cost of providing adequate facilities and buildings is a major problem for governments, corporations, and private organizations. The reasons are many and complex. Facilities and buildings require large amounts of capital, but do not themselves produce goods and services; therefore, decisions to build, lease, buy, renovate, or relocate must be made very carefully.

Construction and land prices continue to climb. Some construction materials and resources remain in short supply. The public calls for reduced taxes. Operations housed in formerly adequate facilities grow more complex and need more space. New laws, energy concerns, a greater interest in worker satisfaction and productivity, and historic preservation movements place added constraints on facilities and make decisions about them more difficult.

As a result, management of space and facilities is becoming more important. Improved skills are required, together with better sources of information and more comprehensive methods for managing facility problems.

Corporate planners, facility committees, and the architects and engineers who assist them labor for months trying to formulate recommendations. Executives, boards of directors, councils, and trustees are frustrated by facility decisions. These groups often depend on building specialists within or outside their own organizations to prepare the facts to make such decisions.

All too often, planners and decision makers focus right away on proposed solutions addressing the question "What should we do?" without resolving a more fundamental question: "What do we need?" They try to solve the problem without defining requirements. Although management decisions about operations are formulated on the basis of carefully prepared facts and figures, many times little analysis is done before facility decisions, which always incur large costs, are made. A financial analysis may be done, but an analysis that establishes a solid definition of needs is often incomplete or too general.

The purpose of this book is to help corporate and government planners, facility committees, and decision makers, as well as the architects, engineers, interior designers, and staff facility specialists who aid them, develop a solid basis for making decisions about facilities and buildings. The book addresses that fundamental question "What do we need?" one which is often overlooked or given cursory attention because of the urgency to find a solution. How to solve facility problems remains the job of design professionals, but solid and complete information about requirements is essential for completing an effective design.

This book attempts to present a simple but systematic approach for defining user requirements for buildings. This approach is comprehensive and suitable for virtually any kind of facility at any stage of planning. It is applicable for small projects in small organizations or large, complex projects in large organizations. In fact, the larger and more complex a project is, the more important the task of accurately defining what is needed. This approach recognizes the need for a computer to manage information on larger projects, and it also recognizes the importance of communicating that information to others, particularly to designers, to ensure that user requirements are applied.

Consequently, this book will be an aid to many people who get involved in planning of facilities. Architects, engineers, interior designers, and other building professionals will find that the method described in the following pages will improve and enhance techniques they have learned or are using, and will help them organize project planning with a client [and obtain essential data from users].

Professionals and semiprofessionals within organizations, such as facility managers, facility planners, space planners, space managers, real property staff, or administrative services staff, can use the method to complete in-house facility planning and management activities. These activities would include validating requests for space or helping organizational units establish the need for space; helping top management decide whether to buy, lease, build, or modify; and preparing to work with in-house or contracted engineering, architectural, or other design services on specific facility projects.

Groups formed by government units, church groups, small companies and organizations, public action groups, and others for a limited or extended period to perform building studies and complete project planning can use this book as the basis for organizing and completing their tasks. This book can be used as a procedural workbook, or merely as a guide, by all of these groups.

The method presented here is based on one that was developed at the U.S. Army Construction Engineering Research Laboratory (CERL) under my leadership for use in facility project planning in the U.S. Army. The original method was structured for just one organization's rules for getting a building defined, funded, designed, and constructed; it has been modified here for general use. The support of many coworkers at CERL must be recognized. Particular credit goes to David L. Dressel, who played a key role throughout the development of the Army method and contributed many ideas to the process. Credit also goes to Martin Koch, who helped with its application to design review. Others who made contributions in the early stages of its development include Wayne Veneklasen, Wolfgang Preiser, and John Burgess.

The method was coordinated with many others and tested on a number of projects. Test projects ranged in size from an office for less than a hundred people to a half-million-square-foot test and evaluation center, and an entire airfield complex. The assistance and contributions of the many people who participated in developing, testing, and applying this method are greatly appreciated. Appreciation is also extended to Robert Shibley and David Lyon, who provided encouragement and support for the development of the original procedure.

Contents

FACILITIES PLANNING

Chapter 1

Introduction to User Requirements

Scenario 1. Although pressed to hold the line on taxes and avoid the ire of constituents, the Facilities Planning Committee of the Hometown County Board faces a state report condemning the county courthouse, which has stood as a community landmark for decades. Their frustration is complicated by their efforts to identify what is needed and how to resolve their facility problem.

Scenario 2. The Fabrication Plant of the Growing Manufacturing Company is bursting at the seams. Sales continue to climb. The company president decides that production facilities cannot be expanded in the existing plant. He assigns his staff to get an expansion program underway. A design firm is hired to solve the facility problem and begins by asking, "What do you need? What are your requirements?"

Scenario 3. The owner of the building where the regional office of Loss Control Insurance Company is located does not want to extend the lease. A new location and new lease are needed. During the past five years in which the current lease was in effect, the operations of the regional office have changed greatly. Both the company and its use of office automation have grown. The automation equipment itself has changed drastically as well, from data cards and one large computer to a multiterminal computer system. Satellite data transmission to the home office is also being considered. The regional office director must know what is really needed for his company so he can look for a new office facility that will be suitable for at least the next five years.

Requirements and Needs

Many more examples pointing to the importance of user requirements in building projects could be cited. Often conditions get so bad that anyone can see that something must be done.

The fundamental question in case after case is "What do we need?" A common answer is to define the solution first. Everyone seems to know how to solve the problem. One person knows about a new building that can be leased. Another thinks a wooded site near the edge of town would make an ideal site for an entirely new building. A third has a friend in real estate who has a building for sale at a "steal" of a price. A fourth has been in a new facility of a similar organization and thinks duplicating that building would be great. Yet whether building deficiencies are large or small, specific needs and requirements must be defined before the best, most precise solution can be determined.

Needs and *requirements* are interchangeable terms. They are statements that define what is expected from a solution, rather than define the solution itself. Requirements do not involve a design or building plan; they are the compilation of statements and data that define what should be. They are not a description of how things are or could be. They must not be confused with a solution.

There are many ways for users of a building to state requirements. One commonly used approach is to identify an existing building as the requirement: "We need a building like [someone else] has." The problem here is that a requirement like that is too vague. If an organization were to move into a building similar to what some other group has, there would most likely be many misfits.

Another approach for defining what is needed is to have every department draw up a floor plan; that is the way one looks at houses. You visit a model home and get a copy of its floor plan or you buy a book of house plans, study them, and decide which one is most appealing and affordable. The problem with user-sketched floor plans is their lack of precision. A designer may have great difficulty in determining what in the floor plan is a requirement and what is not. Moreover, floor plans drawn by different groups obviously won't fit together without major changes.

Another method involves collecting all the data possible and turning these over to a designer, assuming that somehow the requirements will emerge from the volume of data provided. A designer will charge extra to take the necessary time to sort through the data, find out what is relevant and irrelevant, and determine the real requirements. The alternative for a designer is to ignore a great deal of the data, hoping someone will be able to identify inadequacies when solutions are proposed.

A more common method is to hire an architect or other design firm or real estate company to resolve the building problem. Since the people hired are trained professionals, it is often assumed that they will be able to tell a client what is needed; they are expected to prescribe the solution. While a general fit might be possible this way, a building that will work well for a client requires that the designer investigate in detail what is needed. Users must be consulted to obtain requirements and supporting facts. Professionally defined requirements are compiled into a document called an "architectural program" or a "statement of need." The client must pay extra for this service because programming is not a normal part of a designer's fees for a building project, and other organizations must charge a fee for services unless they can recover costs through a later contract, lease, or sale.

Many approaches can be used to define building requirements. Some will be effective; others won't be. What is needed is an orderly process that will establish specifically what users need. Just as important is the presentation of the requirements in a form that effectively communicates them to a designer or others and ensures that they will, in fact, be used and can be accounted for in a final solution.

Requirements of users must be distinguished from *wants;* what is necessary for the building users to accomplish their goals must be distinguished from what would be "nice to have." Unless there is a distinction between requirements and wants, costs may outstrip available funds. The resulting solution may not support user activities after occupancy as it should.

User Requirements

Very often the users of a facility are assumed to be the individuals who occupy or use a building. Sometimes the term *user* refers to the owner or the organization that buys or rents the building, or some individual or body in authority. Whether owner, chief executive, employees, or the public, all parties that have a role in the use of a building are users. This can create confusion when trying to define the user requirements for a building or facility. If there are no conflicts among the requirements of these groups, then no problem exists. However, if the owner or chief executive has different requirements from those who use a building, conflicts can result. At best, a designer can build to satisfy only one set of requirements without compromising other requirements.

Consider an investment company that builds an office building and leases it to a number of firms. Whose requirements are most important—those of the building owner, those of the executive of the leasing firm, or those of the occupying organizations and the workers who make them up? Government buildings are owned by the public, managed by one government agency, and occupied by a second government agency that serves a segment of the public. Whose requirements for the building are most important? What requirements are to be used for design or renovation? Similarly, park and recreation sites are owned by the public, operated and maintained by a government board or agency, and used by the public. How are requirements for these facilities established? Confusion is very common about who is the user and what user requirements form the basis for building projects.

The method described in this book (called the user requirements method or URM) assumes that users are the individuals and groups that will occupy or use a building or facility. Users are those to whom parts of a building are assigned. They are the people who make up each element or unit of an organization: the individuals and groups who occupy workspaces and the people in departments and sections who run operations. Visitors who temporarily use or occupy spaces and people who take advantage of public facilities and spaces are also users. Users may be identifiable individuals we can see and talk to. Users may also be occupants who are yet to be selected.

The user requirements method also assumes that users are goal-oriented and can define the objectives or mission for their activities: in other words, the specific purposes for spaces and features of a given building.

The method assumes that the validity of any user requirements must be evaluated in terms of meeting the objectives or mission of the organizational unit, group, or collection of individuals considered to be the users. How can one justify facility needs? Obviously, one can make general financial arguments for a project about investments, depreciation, and earnings. But the components, such as spaces and features, are justified on the basis of function,—how well the objective or mission of each organizational unit or activity is supported.

The user requirements method assumes that the users of a building know better than anyone else how to perform the activities associated with their objectives or mission. Because the users are the experts, they are the best sources of information for establishing what is expected from the building to support their activities. The method is designed to identify conflicts among requirements of different elements and levels in an organization and resolve them before a facility project is initiated.

From time to time URM assumptions will not be valid and other methods, or modifications to the URM, will be needed. For example, users may not be aware of studies about building utilization rates, marketing data, alternatives for organizational structure, or activities that might improve organizational performance. When users cannot agree on requirements (often the case for shared or public facilities), then techniques for group decision making may be helpful. If a building is being created for an organization that does not yet exist, then user requirements must be developed by those who best foresee its operations. When special cases like these arise, the method in this book can be supplemented by using experts and specialists or applying other approaches helpful in defining requirements. Suggestions for defining requirements for dispersed, future, or ill-defined organizations are discussed in detail in Chapter 3.

For most building projects, the user requirements method is valid. Even if its assumptions are violated, it may still provide a useful approach to defining what is needed.

Who Develops User Requirements?

A variety of people may develop user requirements for buildings. User requirements can be developed by current or future users of buildings if they are trained to perform this task. However, they need a simple method. A variety of building and facilities specialists and professionals can also develop user requirements for current or future users of facilities. The information must come from those who know how operations are or should be conducted, who is involved in them, what equipment is an integral part of these activities, and what time or schedule factors will affect the building. Most often this information can be derived only from those who will be in the building being considered. This data must be converted into statements about the way things should be in a building. These statements are user requirements.

Users may do self-analysis and project the requirements themselves if they know how to proceed and have some aids to assist with key data. Facility-design professionals perform this process regularly and have developed an orderly approach. Their training and experience helps them work with users to develop the statements of expectation. However, the data that drive the user requirements (understanding activities, numbers and kinds of people, kinds and quantities of equipment, and scheduling data about operations) typically come from the users through interviews, group sessions, or survey forms. URM will work for all these groups and may improve techniques already being used.

Why User Requirements Are Necessary

Someone may say, "Why bother developing requirements in detail? It takes too much time. We don't need that much detail to be able to tell someone what we need. It only delays the project. Let's get the project going." Such comments are very common. People are overwhelmed by the urgent prospect of a building project. They want to solve with record speed a problem that may have developed over many years.

There are many possible benefits to be derived from an orderly development of requirements. Even though there may be a short delay in getting a solution started, the benefits could outweigh that delay: money can be saved, a more effective solution can result, the design and construction can move faster, fewer changes during design and construction can result, the solution can remain effective longer, the move-in can be smoother, and users and occupants will feel less upset or threatened by change. Most importantly, because operations do not have to adjust to facility constraints, the cost of operations after occupancy can be reduced.

Cost of Requirements

It takes time to define requirements accurately, but it needn't be a lot of time if an organized procedure is used. With URM, one organization defined detailed requirements for half a million square feet of offices and laboratories in two weeks, and had the data available for query in a computer within 30 days after initial training in URM.

The time or salary of those who participate in formulating requirements must be paid. Regardless of who defines the requirements for a project, some time must be devoted to the task and some expenses will be incurred.

Project Cost

The cost of preparing user requirements can be recovered in a variety of ways. If requirements are accurate, comprehensive, and organized for easy use by a designer, the length of time necessary to complete a design is shortened. A designer spends less time gathering missing information or verifying data supplied, and can move more quickly into developing a solution. The designer will need to make fewer assumptions and can reduce time-consuming interactions with users if requirements are well documented. When reviewing a design at various stages in its development, users can provide more

Figure 1-1. Building development and cost of changes.

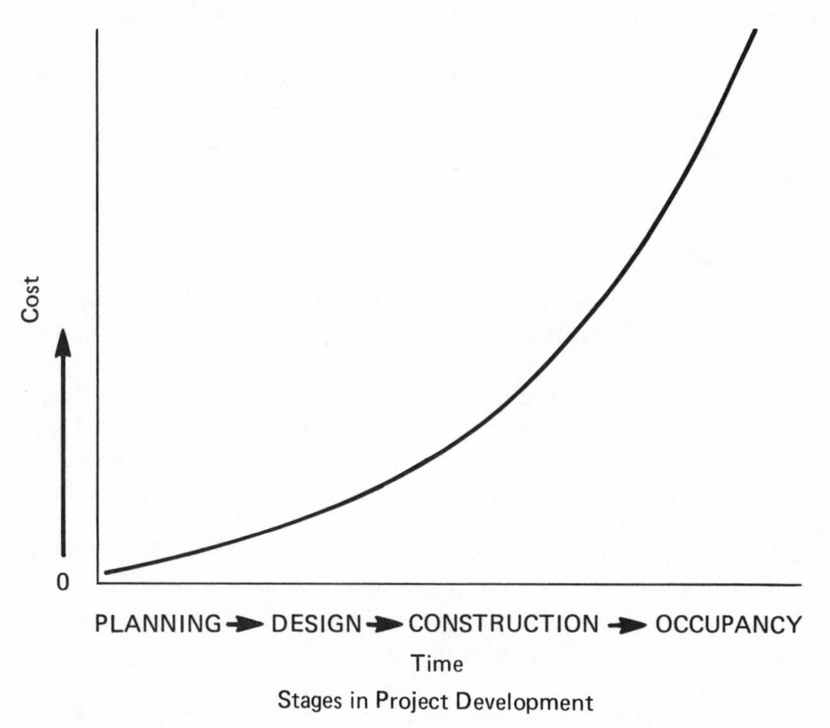

useful feedback to the designers quickly by having well-defined requirements to start with. If users have thought out their requirements well, fewer changes are needed in a design.

Users are often unpleasantly surprised during and after construction by the way things turn out. Each surprise that results in a change or modification adds to the construction cost. And each change delays the project completion. Each change that is not needed but completed will add to operational costs. Each delay on a project can add to the interest cost or affect the value of capital because of inflation.

In general, the further a project moves toward completion, the greater the cost of making a change (see Figure 1-1). It is much cheaper in terms of direct expenses and delays to make changes on the drawing board than during construction. Many changes can be avoided by thoroughly and carefully defining project requirements at the beginning.

Cost of Space

Although other factors are involved, the amount of usable space a project provides is the best indicator of its cost. Whether buying a building or leasing one, having more space than is needed ties up capital that could be used productively elsewhere. No organization is static. It will grow and shrink; its operations will change and its space needs with them. Some excess space is usually needed to allow for organizational change and the reallocation of space that continually occurs. However, it is expensive to get locked into a long-term lease agreement or building procurement that wastes space. One of the corollaries to Murphy's Law says, "An organization will always fill the space available." Excessive space will not always be apparent to the casual observer.

Special features and characteristics also add to the cost of space. Unnecessary features, such as oversized structural, utility, and communication systems or opulent finishes in locations where they are not needed also waste money.

If requirements are accurately prepared, it is easier to make decisions about buying or leasing the correct amount of space, the correct kinds of spaces (office, manufacturing, warehouse, training, et cetera), and getting the right features where they are needed. Unnecessary capital expenditures can easily be avoided.

Cost of Operations

One of the primary purposes of a building is to provide adequate places so that each element of an organization can contribute efficiently toward accomplishment of the overall purpose of its existence. Performing the functions or activities that contribute to the goal costs money. Employees' salaries and expenditures for equipment and materials are expensive. The cost of operations far outweighs the cost of purchase or lease and maintenance of facilities. As shown in Figure 1-2, the National Bureau of Standards estimated for the Public Building Service that the lifetime costs of an office building consist of 2 percent initial building cost, 4 to 6 percent operating and maintenance costs for the building, and 92 percent salaries of the building's occupants.*

Similar results for other kinds of facilities would show that building costs (initial, operating, and maintenance costs) are only a small part of the overall cash flow of a business. The operations inside the buildings create the major cost. Adjusting a building to support operations is generally much cheaper than making operations adjust to a building.

* "Productivity in Civil Engineering and Construction," *Civil Engineering,* February 1983, pp. 60–63.

Figure 1-2. Lifetime costs associated with an office building.

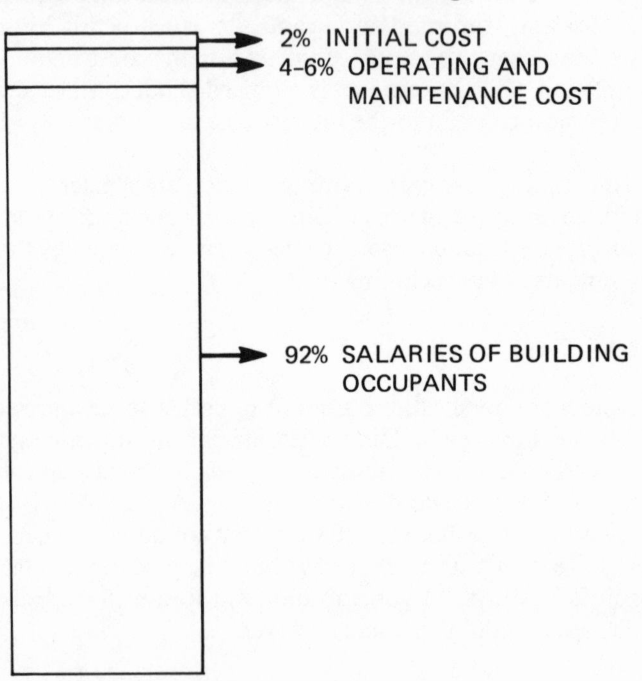

The careful development of building requirements will help achieve a good fit between the building and the users. A good-looking shell is not enough. A good fit is obtained by matching facility needs for particular spaces to the design and features of that space. A good fit will benefit user operations and reduce costs in many ways. Achieving a good fit will enhance and foster occupant performance. Movement of people, materials, and information can be more efficient. People may feel more satisfied with their work and their workplace. While the effect of a good fit between a facility and the operations it contains may be difficult to measure directly in dollars, it can increase productivity and reduce employee turnover or absenteeism.

Cost of Moving

The process of developing facility requirements and the resulting requirements themselves can have indirect benefits when it comes time to move into a new or modified facility. Formulating requirements for buildings requires early planning for new equipment. In URM, existing equipment and furniture are cataloged together with locations. This data will help in planning for and implementing a move.

Fear of Change

Another benefit of URM is reducing the fear and anxiety associated with change. We are in a rapidly changing world. People prefer stability and are apprehensive about change, fearing that it will affect them adversely. Facility changes are just another symbolic or perceived threat. When users are allowed to participate in defining change, they understand the effects better and are more willing to accept the changes that accompany facility projects. URM provides for user participation and can reduce the cost of implementing change.

Costs After Moving In

When people move into a new or modified building, there is a period of adjustment. Obviously, there is a novelty effect to be overcome with any new building. People want to explore and try things out. In a home, for example, the owners will move furniture, pictures, and decorations many times before finding a suitable arrangement. They will adjust furnishings because things don't fit or look right. They will make adjustments to better suit their needs and lifestyle.

Adjustments can be sizable when an organization moves into a building. Deciding how to arrange things, where to put things, and how to get organized should not occur right after moving in. Those decisions should be made during planning. When the formulation of requirements for a building project and the collection of data associated with those requirements are done early, it will stimulate analysis about the transition and provide a foundation for planning a move.

When decisions about operations are delayed until after the move-in, requisitions for new equipment to change partitions, to modify lighting, sound, and thermal conditions, and to make other changes to get things organized are commonplace. Many changes will be turned down because it may not make sense to some people to modify a new bulding or because project funds may already be used up. Yet any money saved by not making necessary adjustments will probably be used up by poor productivity or morale. The occupants are made to adjust rather than having the building fit the needs of the occupants.

By applying the user requirements method, many decisions are made during planning about operations, organization of activities, upgrading of equipment, and other factors. Post move-in crises are avoided and long-term negative effects on occupant performance are avoided.

Summary of Costs

Costs involved in the preparation of requirements can be more than recovered through a variety of direct and indirect benefits. If an organized process for formulating requirements is not used when planning, the requirements will be assumed and decisions associated with them will be made anyway. The costs for an unorganized approach can be high and continue for a long time.

Methods for Defining User Requirements

There are many techniques for defining facility and user requirements. The user requirements method is just one method that can be used. Some methods are specialized for certain kinds of buildings, such as manufacturing plants. Other methods are suitable to particular client-professional relationships. Still other methods require considerable training or experience in order to apply them well. There are methods preferred by particular professionals, such as architects, engineers, or planners. Some of these methods are thoroughly documented in the literature, while others are developed from the experience of the person applying them. The List of References at the end of the book includes some sources that may be helpful for those who wish to supplement the user requirements method.

The user requirements method is suitable for most facility-planning applications. It can be combined with and complement other methods. It gets users involved in change, provides accurate results, and is easily understood. It can be employed by professionals, staff specialists, or those with no training in building planning. It provides users a defensible position, differentiates needs from wants, and is orderly and systematic. It allows users to communicate effectively with designers. It fills a gap found in other methods—that is, how to derive requirements from users. It is easy to use, is suitable for many kinds of buildings, whether large or small, and allows for computer management of data.

Some prefer to use a top-down approach, in which one has a strict set of standards that establishes what kind of space and how much space each individual, organizational element, or activity will get. The standards may include typical features such as privacy, finishes, flooring, utilities, and lighting. This approach assumes that building requirements can be projected from minimal data about users. This approach is satisfactory if a sizable margin of error is acceptable. The standards can be quite useful for planning and early budgeting.

URM is primarily a bottom-up approach. User requirements are derived from actual or projected operating characteristics of organizations; they are summed up to obtain the total. Requirements derived from a bottom-up approach are accurate for operating elements of an organization, but may violate already existing standards. What is important is that requirements derived from a bottom-up approach will provide the best building solution for the operations to be housed. URM can incorporate top-down standards; bottom-up results can be compared to these standards to see if they are out of line. For example, top-down office standards by pay grade, status, and job type can be used in URM. At the same time, if desired, space requirements can be compiled from the activities and equipment of certain jobs to see whether standards are adequate. Violations of existing standards can be evaluated in terms of the overall impact on the organization. Justifications for variances are provided by URM because requirements are directly traced to organizational goals and objectives.

User Requirements and Other Building Data

There are many kinds of requirements for buildings. User requirements provide only some of the data needed for building planning, design, and evaluation. Technical requirements, usually prepared by professionals (engineers, architects, planners, interior designers, facility managers, and others) are also needed. Economic data, marketing data, corporate forecasts, trends in technology, and other data are important as well.

Figure 1-3. Unique project data and general data combined in building design.

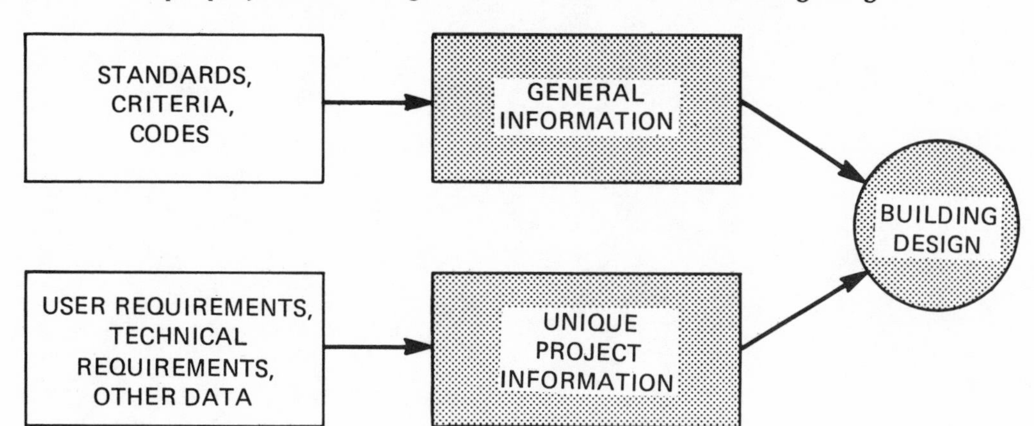

As will become clearer after URM is explored in detail in Chapters 5 through 7, many other kinds of data, besides user requirements used in building planning, depend on or are based on user requirements. User requirements are the cornerstore for project planning and building solutions. Technical requirements such as utility requirements are derived from the equipment, environmental, and other data found in user requirements. Economic analyses and cost studies for a project are not possible or will be inaccurate unless the space and major features of the project have been identified in user requirements. Translating corporate forecasts into operational terms and converting facility deficiencies into projects and budgets is accomplished through user requirements. Even site selection may be heavily influenced by user requirements.

In planning and designing building solutions, professionals depend heavily on their training and experience. Design standards, codes, and criteria are also essential in making a solution effective. Data from these sources provide for a general fit of a building to a specific use. User requirements give the solution a precise fit. One does not go into a shoe store and simply ask for a pair of brown shoes. A customer has other considerations in mind that must be communicated to the sales clerk—dressy or casual appearance, leather or canvas, a particular brand or style, certain type or height of heels, and, of course, a particular size and width. Similarly, with an expensive building purchase or lease, one will use general standards and criteria, but will supplement them with detailed requirements that will make the building most suitable for the organization and the operations that will occur in it. As illustrated in Figure 1-3, a designer will integrate data from both general sources and those that are unique to a project in arriving at a solution. Both sources of data are necessary to achieve a good fit between a building and its users. A further discussion of project data for buildings is found in Appendix D. Several sources in the List of References provide additional data as well.

Summary

The formulation of requirements is an essential part of facility planning. Regardless of who develops requirements or how they are developed, they are derived from users, the organizational units, and people who will occupy and use the facility being planned. Accurately defined requirements are essential for effective decisions about leasing, buying, building a new, or modifying an existing building. If requirements are well defined, the resultant solution can achieve a good fit between building and users. URM is a simple and thorough method for defining requirements and can be used by both professionals and nonprofessionals for virtually any kind of building.

Chapter 2

User Requirements and Applications

In Chapters 5, 6, and 7, the user requirements method (URM) is explained in a step-by-step manner. Before getting to the details of URM, however, an introduction to URM will help one see what is involved and how it works. Understanding URM provides a sound basis for its application to a variety of building problems. A number of variations for the method are possible depending on the kind of application, type of building, type of user group, number of people implementing it for a project, experience and training of those using the method, and other factors. Variations suitable for different applications are discussed later in Chapter 8.

A Brief Look at URM

The user requirements method is a process for defining user requirements for a building or facility. It is not a method for developing layouts or designing a solution. It is a process for systematically formulating, justifying, and communicating what is needed by building users so they can perform activities efficiently and effectively after they move in. The compilation of user requirements that results from this method provides the basis for making decisions and achieving solutions. The method does not provide all the data necessary to support building decisions and projects. Technical, economic, political, and other data are used in decision making and designs for buildings as well as functional requirements of users.

The Process

There are three major steps in the user requirements method, as illustrated in Figure 2-1. In Step 1, the operations and activities of users are analyzed: who will be in the building, what it will contain, and how it will be used. In Step 2, the user requirements for the facility are defined and recorded. In Step 3, the data from Steps 1 and 2 are organized and documented in a form that makes it easy for designers to understand and use. This is necessary because the thought processes users follow in compiling requirements are different from the ways designers of solutions look at the data. Users think in terms of activities and organizational units, while designers think in terms of room, types of space, and similar units of space. Much of this last step is clerical and would benefit from computer assistance.

Figure 2-1. The three steps in the user requirements method.

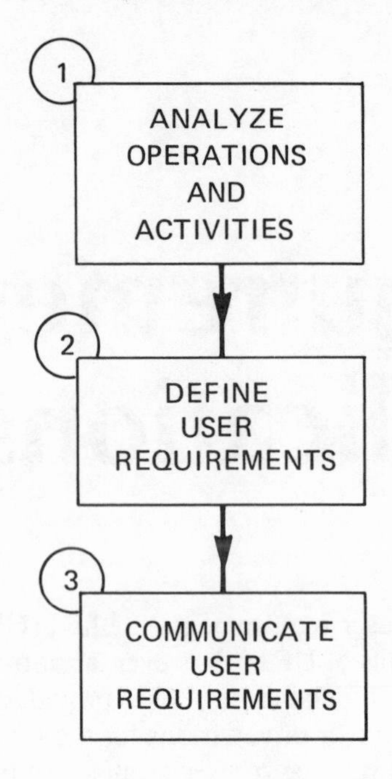

Typical Staffing

User requirements and supporting data are derived from building users. Because no single individual knows all the details about how an organization or group of people operates, one must derive requirements from each organizational unit. In URM, a spokesperson is required for each department, division, section, branch, or other organizational unit. These individuals, called *representatives* in URM, are the key sources of data for their respective organizational units.

Typically, someone is in charge of the URM process as well. One person is the *coordinator* and serves to keep URM moving, resolving procedural and substantive problems, and ensuring that the final product, user requirements, is compiled. The coordinator may get assistance from others.

This staffing structure assumes that users will develop their own requirements. However, roles of coordinator and representatives may be handled by professionals or specialists in facility planning. A variety of role adjustments is possible when special staff or contracted professionals get involved or when special applications of URM require some adjustments in roles.

Features of URM

URM has many features which make it advantageous:

1. *It is easy to apply.* URM starts with information users know. It progresses in a step-by-step, logical manner. Little training is required to apply it.
2. *It gives defensible, objective results.* All analysis is derived from the purpose for an organization's existence. Needs are distinguished from wants.
3. *It is based on a simple conceptual model.* URM users first learn how their activities, equipment, people, and time are related to buildings and cost.

4. *It has many applications.* URM is valuable for evaluating existing buildings to identify problems before they become severe, planning new projects, communicating to designers what is needed, reviewing designs and evaluating candidate buildings for purchase or lease, creating standards, and other applications.
5. *It allows for flexible staffing.* URM can be performed by building occupants themselves, staff facility specialists, or hired professional designers and planners. The workload can be distributed or URM can be implemented by one or a few persons.
6. *It aids continuity and change.* URM aids continuity by documenting requirements so that one individual is not the only source of information. URM aids change by getting users involved early in planning for change.
7. *It is accurate.* URM provides very accurate information about what is needed. The method can be adjusted so that the appropriate degree of accuracy is developed.
8. *It achieves quality.* URM does more than estimate how much space of various types is needed. It also focuses on qualitative details that are important for productivity of building occupants.
9. *It helps control cost.* Because URM can define requirements accurately, project costs will be on target. Unnecessary features are easily identified.
10. *It is an effective communication to designers.* Data from URM are organized so that both users and designers can find particular requirements easily. Because of this, data are more likely to be used in developing a design.
11. *It is designed for automation.* Data elements resulting from URM are standardized so that available data-base management systems can easily be used. Special computer programs need not be written.

Users of URM

Facility users will find URM a logical, orderly process that begins with information users know well. The process is composed of small increments, so that one is not overwhelmed with complicated procedures. Worksheets, examples, and other aids are provided.

Architects, engineers, interior designers, and other planning and design professionals can apply URM in their professional practices. Many will find that URM improves their current programming methods. It is a simple approach to building planning and programming. It yields comprehensive data with a minimum application of manpower. Data can be easily collected from organizational units or much of the effort can be transferred to others to minimize the time required when planning staff is limited.

URM can be applied by virtually anyone. The amount of training in building technology one has is not an essential criterion for establishing whether the method can be completed. Building users with no training in building planning and design, building design professionals whose training was devoted to building planning and design, and staff facility specialists with varied amounts of training can all use URM. Obviously, those with training and experience will be able to learn URM more quickly and find shortcuts for its use.

The Building Cycle

There are several ways to obtain a building: build a new or buy, lease, modify, or expand an existing building. When building a new, there are a number of options: build a premanufactured building, design and build in a normal sequence, fast-track (design and build in parallel), and other approaches.

Regardless of the method, a facility project begins with requirements. Similarly, when leasing, buying, or adjusting an existing building, the planners must begin with requirements.

Figure 2-2 illustrates a general process for new construction. URM is helpful in varying degrees at each step. The building cycle begins with a recognition of a facility problem. Planning and budgeting activities are then initiated. When an approach for a solution is selected, detailed project planning and data collection is conducted during predesign. The project design is then completed and construction begins. While the building is being prepared, equipment to be placed in the new facility is procured. Finally, users move in. The loop is completed as new facility problems are recognized during occupancy. A similar model could be drawn for each method of obtaining a building or building solution. Regardless of the method for getting a solution, the process is an ongoing, cyclical one.

Recognizing Facility Problems

Facility problems are often recognized only when they reach crisis proportions, usually because they build up over time. Some are resolved through maintenance. Others are tolerated until the effects are so severe that everyone agrees that something must be done. URM can help identify functional problems early. By comparing the way things are to what they should be, deficiencies are easily identified. It is easy to see what one has; getting requirements defined is always the difficult part.

Planning and Budgeting

Once a facility problem is recognized, the next questions are "What can we do about it?" and "How much will it cost to solve it?" General estimates about what is required, the alternatives, and their costs are evaluated through feasibility studies. Estimates will include building costs, new equipment purchases, and moving expenses. Budget requests are introduced to allow for initial financial planning.

The three key factors in this early phase are the scope of the project (how large), possible solutions, and cost. User requirements are needed during this phase. However, only enough information to support the three key factors is collected at this stage.

Predesign

This can be considered an extension of planning and budgeting. The main difference is detail. This phase begins when an approval to move toward a solution is granted. Both user requirements and technical requirements are defined in more detail than in previous steps. These data provide the basis for design or evaluation of specific buildings that are candidates for purchase or lease.

Design

In this phase, a design is developed and evaluated against requirements. If the purchase or lease of existing facilities is considered, candidate buildings are identified and evaluated.

Construction or Modification

During this phase the design is implemented. Similarly, a lease or purchase agreement is executed. User requirements do not have a significant direct role in this phase. They have been converted into drawings and specifications that are used to ensure that requirements will be met.

Figure 2-2. Typical steps in the development of a new building.

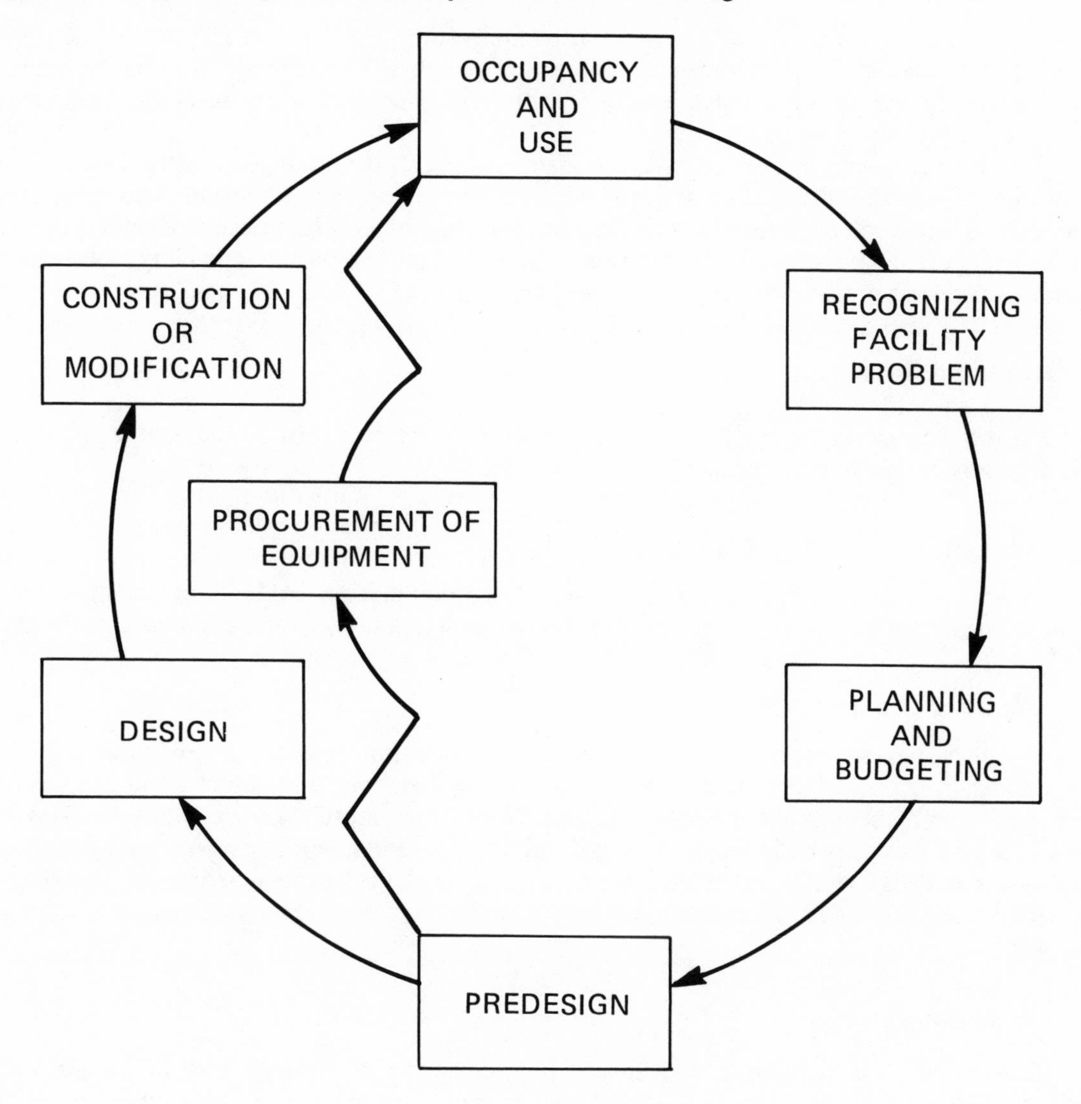

Procurement of Equipment

Concurrent with the design or selection of a facility, new equipment is ordered so that it can be installed at the right time. The need for new equipment is usually identified during the planning and budgeting and predesign phases. URM will help identify the numbers and kinds of equipment needed, because equipment to be brought into a facility is tabulated as part of URM.

Occupancy and Use

Initially, operations are moved in and started up. As time goes on, new facility problems arise because of changes in the organization, equipment, or activities. These changes may be the result of many things, including technological, social, economic, or other factors that affect the organization and its structure, objectives, and activities.

User requirements must be continually updated in light of these changes to detect facility problems early and aid in formulating and achieving solutions. Problems do not have to reach crisis proportions before solutions are sought. A program of continual modification and upgrading to meet the demands of change is needed. This continual process of identifying occupant-facility problems and gaining solutions has recently been called *facility management.*

Leasing and Buying

The process for leasing or buying facilities is quite similar to the new construction process. Figure 2-3 illustrates the lease-or-buy process.

Recognizing Facility Problems

As with new construction, one begins with a recognition that current buildings are inadequate. Comparing requirements to existing facilities can pinpoint quantitative and qualitative problems.

Planning and Budgeting

Again, the next step is planning and budgeting. The approximate size and type of building needed must be known. These data can be derived from user requirements. If the purchase of a building is being considered, the current market values of suitable buildings must be known. Many variables will affect the cost of existing buildings, such as age, condition, location, type of construction, size of site, and access to roads and railroads. If leasing is being considered, the lease rates in various locations for suitable buildings must also be known: costs per square foot or annual rental rates per square foot are needed.

Prelease or Prebuy

Just as with a predesign for new construction, detailed requirements must be known. After general approval is given for seeking a solution, a precise estimate of types and amounts of spaces and kinds of features is developed so that particular buildings available for lease or purchase can be looked at.

Lease-or-Buy Options

In this step, candidate buildings are located. Each is evaluated for cost, location, ability to meet requirements, the need for modifications (and their cost), and other factors. As in comparing require-

Figure 2-3. Typical steps in leasing a building.

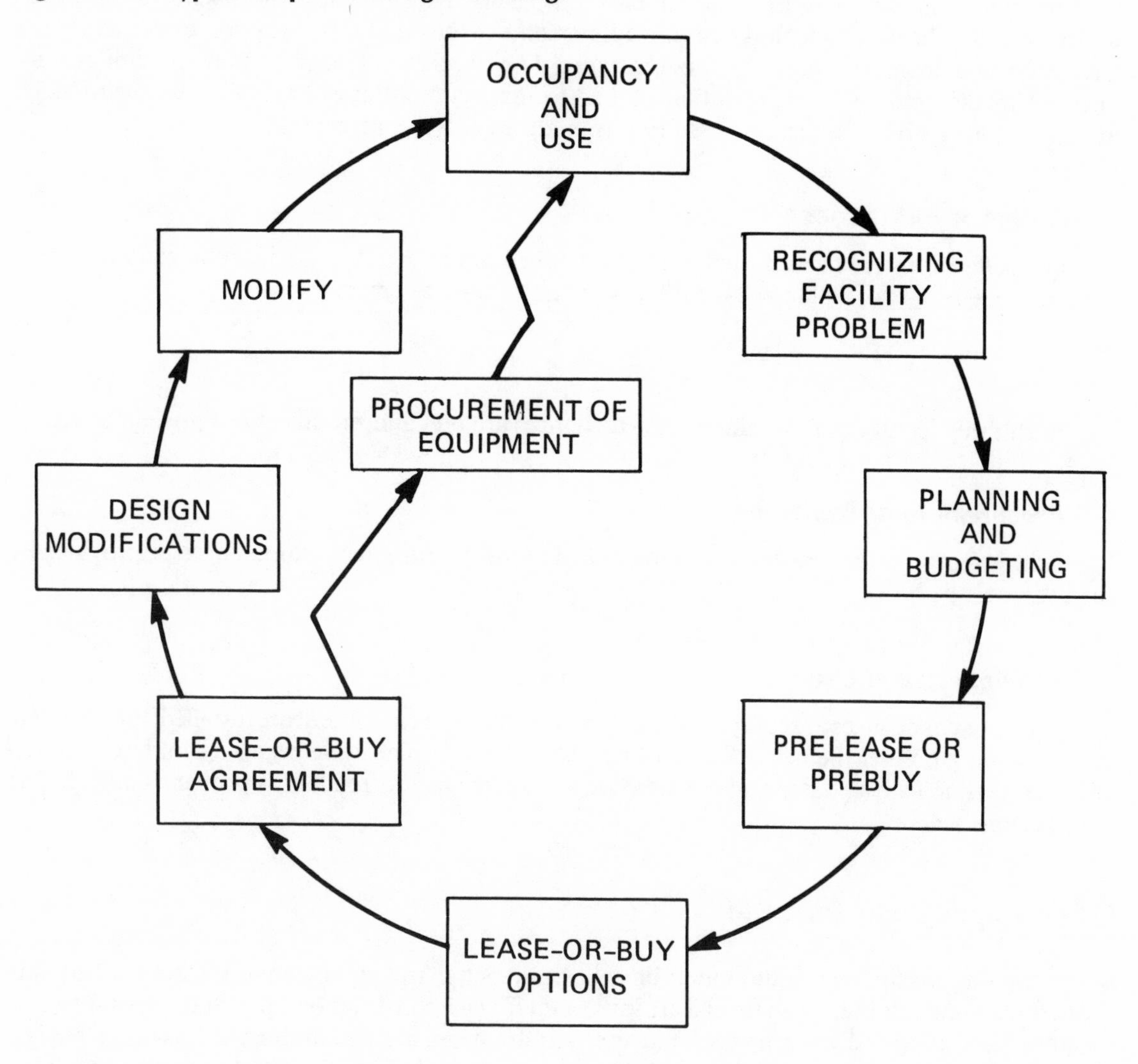

ments with existing buildings to identify problems with them, each candidate building or group of buildings can be evaluated by comparing size and features with requirements to establish how closely they satisfy requirements. Results of the evaluation for each candidate can be compared to establish which one is best.

Lease-or-Buy Agreement

After a best candidate facility is determined, an agreement to lease or buy must be executed. No existing building is likely to satisfy requirements completely. Therefore, lease arrangements may need to stipulate that the lessor will make certain adjustments to the building or its price, or that the lessee has permission to make necessary modifications. The purchase agreement may be similarly influenced by the amount of modification needed to make a building suitable for use. Deficiencies identified during evaluation can be used to formulate an effective agreement.

Design Modifications

If modifications are needed, they must be planned out in detail. Even if modifications are not needed or are minimal, a plan or layout for occupancy must be created.

Modify

In this step, contracts for making needed changes are met and modifications are completed.

Procurement of Equipment

Concurrent with the design and completion of modifications, new equipment to be provided by the user must be procured.

Occupancy and Use

Finally, after changes are completed and new furniture and equipment are installed, the organization moves in and uses the newly leased or purchased facility. The cycle begins again. User requirements are periodically updated and compared to the current facility to identify problems early, before they become severe.

Other Applications for URM

Many applications for user requirements have been alluded to in the previous discussion. URM and the user requirements that result from it are important for new construction projects or explaining to a designer what is needed for a project, but user requirements are not limited to leasing or buying buildings. They are essential in facilities planning: they provide the basis for detecting facility problems and evaluating their severity. User requirements are an essential element in early planning and budgeting and in feasibility studies: they provide a foundation for projecting the size and cost of a building and for exploring alternate solutions. User requirements are obviously needed to explain to a designer what a facility project should provide; they also provide the basis for users to evaluate and review designs and buildings that are candidates for purchase or lease. User requirements contain data that will aid moving in or relocating. By getting building users involved in change, the URM process can help reduce the anxiety associated with changes in facilities.

While each of these applications may need slightly different information or details, only slight variations to URM are needed for each application. Particular applications will be discussed in detail in Chapter 8 after fundamentals of URM are presented in Chapters 5, 6, and 7.

Summary

The user requirements method is an easy-to-use method for developing user-functional requirements for buildings. It is completed in three major steps. URM can be used by architects, engineers, interior designers, facility managers, facility committees, space managers, administrative staff with responsibility for facilities, and even by users who have no training or experience in building planning. User requirements that result from URM complement technical requirements and other data important in facility planning. Whether new construction, leasing, purchase, modification, or other facility changes are anticipated, URM can play an important role in achieving facilities that meet the needs of occupants.

Chapter 3

How to Use This Book

There are many kinds of requirements that affect building projects. They include user, technical, regulatory, safety, legal, economic, and political requirements. Successful preparation of all requirements for a building project depends on many kinds of people. These include marketing analysts, corporate planners, administrators, executives, occupants and users (employees, customers, the public, even maintenance people), planners, architects, engineers, owners, and facility managers. Each person has a role in defining some of the requirements. In part the roles reflect the kinds of requirements needed for facilities. Seldom can one person be the source for all requirements.

User requirements are one of the two well-recognized kinds of requirements for buildings. The second is technical requirements. Engineers, architects, planners, and other design professionals deal primarily with the technical requirements of a project. Users have the primary role in defining user requirements, even when design professionals assist them.

Often one assumes that the boss or someone in authority knows best what is needed. However, top management doesn't know how to accomplish the detailed tasks of its organization as well as the individuals at the lower end of the organizational structure who perform them. Each level and unit in an organization has a role in URM and can contribute to the development of user requirements. Someone will supervise the mechanics of the process. Someone from each organizational unit will serve as a representative. But everyone in an organizational unit can review and provide input on draft requirements and supporting data before they are made final. In Chapter 4, recommendations are made for staffing URM to get users involved in preparing requirements.

While this book is written with user organizations in mind, the method is also very useful for others who must work with or in place of users on building projects. Building-technology professionals, such as architects, engineers, or interior designers, who are hired from outside the organization that has the building project, can apply URM. Staff-facility specialists from within the organization, such as facilities planners and managers who may be professionals or nonprofessionals in building technology, can apply URM as well. Large organizations or small ones will find URM helpful. Whether leasing, buying, building new, or modifying existing buildings, URM will be effective.

This chapter will summarize how different participants in building planning can use this book in different situations.

Users of URM

Regardless of who uses URM, whether user or nonuser, professional or nonprofessional, large organization or small, one should become familiar with the development of user requirements as described in detail in Chapters 5, 6, and 7. Whatever kinds of facility problems URM will be applied to, the forms in Appendix A will be helpful. If a computer will be used to manage data, Appendix B will provide ideas for structuring data-base files. No matter who uses URM, this book can be used as a reference for each participant.

User Organizations

Organizations that intend to apply URM on their own will find Chapter 4 very helpful. It explains how to get organized and prepared for URM, and suggests how to staff URM so that tasks are distributed and no one person is overburdened.

The coordinator, who leads the URM application, must have a thorough knowledge of URM. Various aids and examples are scattered throughout Chapters 4 through 7. If the coordinator needs to know about facility project planning in general, Appendix D explains what data may be needed in addition to user requirements. The List of References will also be helpful in locating general information on project planning. The coordinator will find ideas in Chapter 8 for adjusting URM to various kinds of facility projects.

Architects, Engineers, and Other Designers

Architectural, engineering, interior design, and related professional firms that provide facilities planning and design services to clients will find this book helpful. Most architectural and engineering firms use some method now for facilities planning and architectural programming. URM can become a major portion of architectural programming activities. Portions of URM can be extracted from Chapters 5, 6, and 7 and used to modify, supplement, or improve architectural programming procedures currently in use. URM might also replace present approaches.

In Chapter 4, there are suggestions for planning and design professionals who manage and execute URM for client organizations. Planning and design professionals can assume several roles in the URM process, particularly that of coordinator and part or sometimes all of the representatives' roles. The suggested adjustments to URM in Chapter 8 for various kinds of facilities projects are very important for these professionals. Also, the ideas for managing URM data with computers will be of interest.

Corporate Facility Departments

Large companies and organizations usually have a staff element responsible for facilities planning and management. Internal procedures are developed and used for identifying facilities problems and implementing facilities changes across the entire company.

People in the organizational units responsible for these functions will find URM valuable. URM procedures can be used in at least two ways. Major elements of an organization that think they have facility problems, and are seeking a new or modified building, can be assigned to define their requirements. This book can serve as their procedural manual. An alternative procedure is to help these elements develop requirements. The facilities department can work much like an outside firm and assume the role of the coordinator or, perhaps, those of representatives of organizational units. The recommendations for planning and design firms are applicable to the corporate facilities department as well.

For either way of dealing with facility problems, the URM process, or portions of it adjusted to fit into already existing procedures, can be used to improve, modify, or replace current requirements-development methods. The ideas in this book should provide ways to involve users in converting organizational change into facility change. Not only are Chapters 4 through 7 important, but Chapter 8 will provide ideas for incorporating URM into current procedures for dealing with a variety of facilities problems.

Facility and Space Planners and Managers

Some small organizations or companies, as well as local elements of large organizations (such as branch offices, plants, or terminals) assign the responsibility of dealing with facility and space problems to someone on the administrative staff. Often these individuals do not have training in architecture, engineering, or other building-related specialties. These staff specialists will find URM very helpful in getting facility problems defined and justification for actions prepared.

Again, knowledge of URM in general (found in Chapters 5 through 7), how to staff requirements development, and get URM organized (Chapter 4) are essential. A review of various applications of URM in Chapter 8 will help adjust steps to meet particular facility planning activities. In space management, user organizations may find URM to be the best method for defining and justifying their requests for more or different space.

Planning Committees

Some organizations, such as small companies, school districts, local governments, churches, and private organizations, face facility problems infrequently. When facility projects or potential projects do come up, it is common to form a committee to help decide what to do. The members of the committee may be respected, responsible individuals, but probably will not have training or experience in facilities planning. They may have knowledge of how buildings are constructed, but that will not help much during planning. These organizations typically do not have procedures in place for dealing with facility problems. For such groups, URM can be very helpful because it provides an organized way to get started by defining what is needed. With a full knowledge of requirements, the committee can evaluate how bad the facility problem is and what actions will best meet the needs at a reasonable cost.

Knowledge of URM basics in Chapters 5 through 7 are important here. Staffing may vary somewhat from that described in Chapter 4. The committee will often take on the role of coordinator and representatives may come from each functional unit of the organization. A formal structure may not exist for the organization. In some cases it is helpful to identify the major functions of the organization and the activities within each function. The activities may be the only kind of recognized elements, with the same members of the organization participating in different activities at different times.

The committee can refer to Chapter 8 to see what actions to resolve facility problems (such as problem definition, design review, et cetera) might be required and how to use URM to complete the actions.

Using URM

Through URM, data is compiled that can be used in a number of ways. This book provides help with these applications, particularly in Chapter 8. Before turning to these different applications, a brief discussion of what URM really does is in order.

URM is a method for defining and documenting how things *should be* or *ought to be.* URM data can be compared to the *way things are* or *will be.* As illustrated in Figure 3-1,when this comparison is made, differences become apparent. These differences may be problems, excesses, shortages, or deficiencies. Differences may be good or bad. They may or may not require adjustments or corrective action.

This comparative approach offers several uses for URM. If the subject in question is a current building, requirements can help identify the problems to be resolved. They may be quantitative (space

Figure 3-1. Differences are detected when requirements are compared to existing or future facilities.

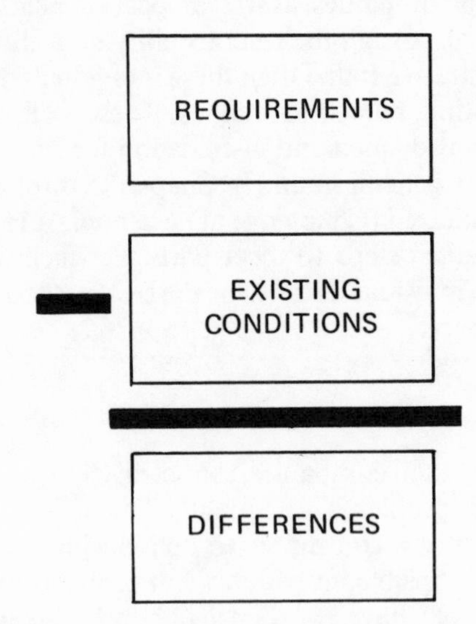

shortages or excesses) or qualitative (missing or inadequate features). URM data become the standard that existing buildings and spaces must meet.

URM data can be compared with the drawings and specifications of potential buildings. One can compare requirements with these drawings and specifications to see if the design does satisfy requirements. The design can represent new buildings or modifications to existing ones. If requirements are not met, the necessary adjustments to the design can be identified and made.

Potential buildings can be candidates for lease or purchase. Each candidate can be evaluated against requirements. Each potential building can be evaluated against URM data. The differences between requirements and each potential building can be noted systematically. Results for each can be compared to see which produces the fewest differences or those that are the least costly to correct so that requirements are met.

Some uses for user requirements require greater accuracy in data or more details than others. As a result URM must be adjusted slightly to fit particular applications. Early planning generally requires less data and less accuracy than does predesign, actual solutions, and move-in. For example, move-in requires precise data about who goes where and what equipment is moved from old locations to new locations. URM data, originally developed for planning and design, must be supplemented to prepare for move-in. By using URM, however, much of the data were previously compiled and need not be collected again.

In Chapter 8 readers will find specific recommendations for using and adjusting URM for various applications.

Whether a facility project involves a) leasing, buying, or building, b) small, medium, or large organizations and building projects, c) organizations that are scattered at different sites, or d) ill-defined and future organizations, URM will provide help in defining what is needed.

Build, Lease, or Buy

This book is useful whether one is considering leasing a building or a portion of one, buying a building, building a new one, or modifying an existing one. Regardless of what solution to a building problem is selected, each begins with a statement of what is needed. URM doesn't have to be different for each kind of solution. Any potential solution can be compared to requirements.

Small, Medium, and Large Projects

URM is the same no matter how large or small a project may be. The number of organizational units involved or who serves as coordinator or representatives may differ, but the procedure is the same for small, medium, and large projects. Since URM is applied independently by individual organizational units for most of the process, and results are combined to form the total set of requirements, URM is not adjusted for the size of a project.

Because data compilation is distributed to representatives, the time required to prepare data will not differ from large to small buildings.

Because URM is not changed for small, medium, or large projects or for the kind of organization applying it, the examples in Chapters 5 through 7 are applicable to any size of organization. The examples are for organizational units or to illustrate procedures that are applied to any size of project. Lists may be longer for large projects, but the format and content will be the same as for small projects.

Scattered Sites

Sometimes when a facility project is being considered, the goal is to consolidate several organizations into one location to improve operational effectiveness and reduce cost. Redundant activities or organizational units may be eliminated or reduced through consolidation. In this situation URM itself needs little or no change. The coordinator may have to travel to each site and conduct training for representatives (see Chapter 4), but the procedures are the same at each site.

It may be necessary to have an assistant coordinator at each site who can perform locally some of the functions of the coordinator. These functions would include serving as a point of contact for procedural questions, collecting data forms from representatives, and aiding in communication between the coordinator and organizational units within the site. Assistant coordinators can stay in touch with the overall coordinator by phone and mail.

Another adjustment in using URM on remote sites is coordination meetings. Near the end of URM assistant coordinators or representatives may have to get together at one site to identify overall relationships.

Future Organizations

URM gets people in existing organizations involved in planning for change. But what if the organization being planned doesn't exist? Can URM be used? Data requirements are the same whether an organization exists in reality or simply on paper. Someone must provide data for the theoretical organizational units. URM can be very helpful when dealing with future organizations because it forces the people who are substituting as representatives to estimate details that are important and may otherwise be overlooked. Accuracy may be reduced, but need not be if operations are well thought out and researched.

Who can make projections for a future organization? The best approach is to find those who are most knowledgable. Several people might be involved. Representatives may be architects, engineers, or others from inside or outside the organization, but they must be knowledgable about the people, activities, equipment, and time that drives the user requirements. One person might represent several organizational units and seek answers to provide URM data. One can use similar organizational units in existing organizations to estimate the requirements for future organizational units. In some cases experts will have to be hired to forecast requirements. Each application will be different. But URM does not need to be changed, only the roles and participants. For dealing with future organizations, one should thoroughly understand the staffing possibilities (Chapter 4) for URM, as well as the process (Chapters 5 through 7).

Ill-Defined Operations

Sometimes an organization is being created or modified, but it has not been determined exactly what organizational units will exist. Sometimes the operations of an organization have not been fully defined, or one of two or more approaches has not yet been selected. How can user requirements be developed for ill-defined organizations or operations?

In general, the operations and structure of an organization must be known before a precise estimate of requirements can be completed. However, more is often known than one thinks. If the difficulty stems from the fact that operations have not been fully developed, then one will have to make general estimates of requirements or wait until details of operations are resolved. If it stems from not having decided which of two or more alternatives for operations will be selected, then requirements may need to be developed for each alternative. URM can proceed as discussed in Chapters 4 through 7, perhaps with a few adjustments for each alternative. Further recommendations are found in Chapter 8.

Summary

All participants in facilities planning will find this book helpful in defining user requirements, whether they are building users, design professionals, staff specialists, or experts who assist. The approach for each user of this book is fundamentally the same, although roles and responsibilities may vary. All need to be familiar with the general URM process found in Chapters 5, 6, and 7. For special applications, adjustments to URM are appropriate for efficiency and proper level of precision. Recommendations for these adjustments are found in Chapter 8. The responsibilities and roles of user coordinators and representatives of organizational units are found in Chapter 4, which also discusses ways others can participate in or assume these roles.

Chapter 4

Getting Ready to Apply URM

A building has an important relationship to its users. To some, a building is merely a place. But buildings are more than places. They need to support activities that occur in them. The characteristics and features of a building can influence how well or how efficiently the activities are completed.

Six Keys of Organizational Accomplishment

The Six Keys of Organizational Accomplishment give building users a conceptual framework for dealing with the relationships between a building and the activities it is intended to support. Every organization, no matter how formally it is structured, has a purpose for its existence. Each organization has some goal it seeks to accomplish. Six keys, working together, are necessary to open the door to accomplishing organizational goals. As noted in Figure 4-1, these six keys include 1) *activities* of the organization, 2) *people* or personnel to perform the activities, 3) *equipment* and supplies to make the activities feasible, easy, and safe to perform, 4) time or *schedule* to complete the activities, 5) a *building,* facility or place in which to perform the activities, and 6) *funds* to pay for the other five keys.

The PEAS of URM

The six keys concept is the basis for the user requirements method. The concept provides a strategy by which users can define what they need in a building. Users know the most about the first four keys in the concept: their *people or personnel, equipment, activities,* and *schedule.* These four keys are called the PEAS of URM. They are illustrated in Figure 4-2. By analyzing these elements and deciding which will be in or affect a facility, users have a solid basis for defining what characteristics and features must be provided by the building (fifth key) to make the organization successful. Then, one can estimate the necessary cost or funds.

The user requirements method helps bridge the information gap between users and designers and ensure successful building solutions. URM allows users to translate their operations into statements about what is expected from a building. Developing an effective building solution is left to the professionals.

Figure 4-1. The six keys of organizational accomplishment.

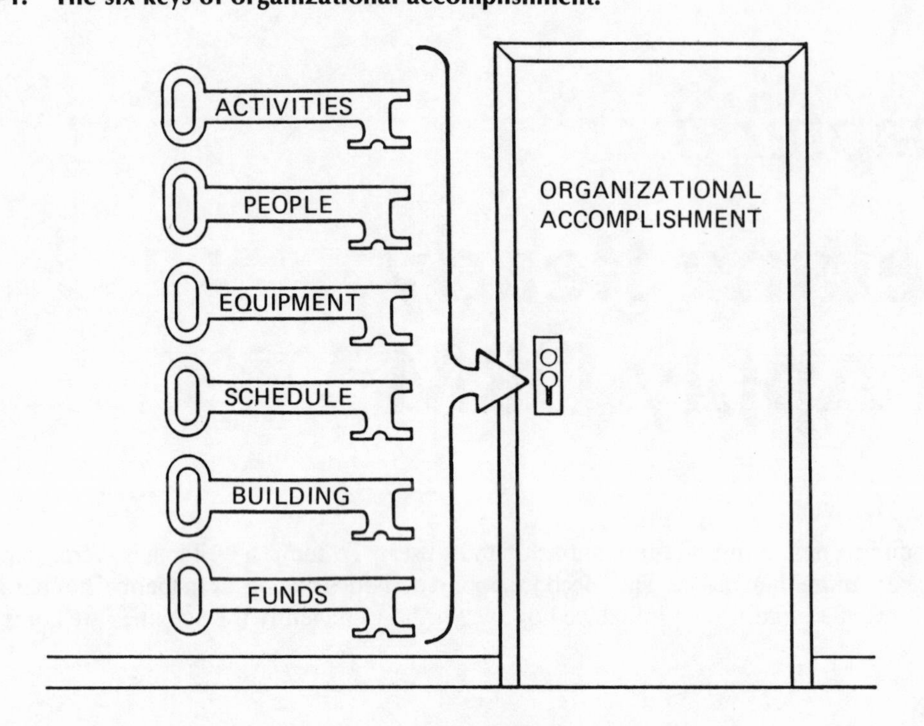

Figure 4-2. Four keys form the basis of user requirements for buildings.

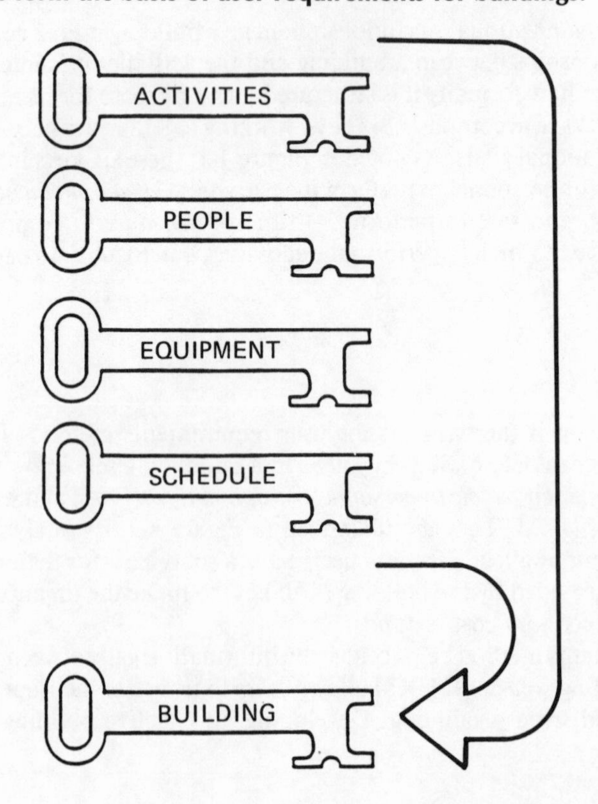

Staffing URM

Like any other activity, URM is easier to do with a little planning. One must decide who will do what and when. Implementing URM is not a trivial task. Good leadership is required to complete assignments on time and to avoid delays in a building project. Many people are likely to be involved and need to know what to do. This section offers suggestions for staffing URM.

General Suggestions for Staffing

URM requires the participation of many people in an organization. As noted in Figure 4-3, it requires someone at the top of an organization with authority, a coordinator or leader, the head of each organizational unit, and a representative from each organizational unit. It can also involve many other individuals in the organization. Each participant has a different role, but each role is important.

Figure 4-3 also identifies where staff facility specialists and design professionals fit in. They can assume some of these roles. For example, they can facilitate and manage URM and execute many URM tasks. They can serve as coordinator or leader, collect and record data from organizational unit representatives, help resolve conflicts, and prepare final documentation. However, data must come from users, and decisions about requirements must be made by users. Staff facility specialists and design professionals can suggest, recommend, assist, aid, and coordinate in this bottom-up approach. As soon as staff facility specialists and design professionals authorize, approve, and assign user requirements, a top-down approach is in action.

Top Management

The individual in charge of an entire organization or someone in authority at the highest level has three important functions in URM. First, URM and the project for which it is used must be given high priority in the organization. Everyone must understand that developing user requirements is very important. Approval for assigning resources and people to URM must be given. It must also be clear that progress toward completion of user requirements will be monitored.

Second, top management must provide high-level information about the organization and its goals. Not only does this include estimates of growth and direction for the organization that are typically prepared by the corporate planning staff, but also must include requirements for the building in general or requirements that may affect some or all organizational units. As projections are made about the future in terms of the four PEAS of URM (people, equipment, activities, and schedules of organizational units), heads generally turn upward in an organization for help in projecting change.

Third, top management must serve as a judge or arbitrator about conflicts among organizational units concerning user requirements and the activities, personnel, equipment, spaces, and other information involved in the conflicts. When staff facility specialists and design professionals are managing URM, they can help identify conflicts. They can seek solutions to conflicts at appropriate management levels. One may not have to go all the way to the top. Generally, conflicts can be resolved at one level in the organizational structure above those units involved in the conflict.

URM Coordinator

Someone must be placed in charge of implementing URM. The success of URM depends heavily on this person, who will be called a *coordinator.* The coordinator is selected by top management, given the authority to perform all the functions of a coordinator, and authorized necessary staff. Without full authority, the coordinator cannot be effective and URM or any other planning method will fail. Also

Figure 4-3. Participants and roles in URM.

Participants	Roles
Top management	—Give URM high priority —Source of information —Resolve conflicts*
Coordinator*	—Expert in URM —Implement URM —Train URM participants —Collect data from representatives —Complete documentation
Representatives*	—Prepare data for organizational unit
Head of organizational unit	—Review data for organizational unit
Others in organizational unit	—Verify data —Suggest improvements in data

*indicates where staff facility specialists and design professionals can assume roles in URM

at this time, who the coordinator reports to should be established. Normally, the coordinator works directly for a very high level of management, a president, vice president, director, superintendant, board of directors, or other top official. The coordinator is selected by the person to whom he/she must report. The selection may be made from candidates provided by departments or from the top management staff. An example of a memorandum by which a coordinator is appointed is provided in Figure 4-4.

Because the URM coordinator's role is so important, selection of the individual for this job must be done carefully. It is very important that the coordinator have leadership qualities. He or she must be able to manage and motivate other people and solve conflicts. The coordinator must be skilled in working with all levels of the organization. Communication by the coordinator up and down the organization is vital. The coordinator should have a general knowledge of all operations and activities of the organization. In-depth knowledge is not required, but would be an asset. Some knowledge and experience about building planning, budgeting, and construction would certainly be helpful. Because the coordinator is a central figure for a building project and provides continuity to facility improvement programs for the users, consideration must be given to how long the individual can serve or is likely to serve as coordinator. Coordinator characteristics are summarized in Figure 4-5.

The coordinator of URM performs several functions. He or she serves as the expert in the URM process, plans and schedules URM activities, secures representatives, prepares some URM data, prepares final URM documents, conducts major URM meetings, and serves as the liaison for URM data. The key functions of the coordinator are summarized in Figure 4-6.

Serves as the Expert for the URM Process. Someone must be knowledgable about the URM process and what steps need to be accomplished next. Someone must be available to give advice about procedures, use of forms, and how to handle certain kinds of data in a consistent manner. The coordinator, as the one most knowledgable about URM, is the person people call on for help.

Organizes URM Activities. The coordinator sets a schedule for completing URM activities, publishes the schedule, and keeps participants informed about what needs to be done when. The coordinator also calls most of the meetings, handles meeting agendas, arranges for compilation of data provided by representatives, watches for problems in completing URM activities, and assists in securing solutions for them. An important role of the coordinator is communication—keeping participants informed about what is happening in the process.

Trains URM Participants. The coordinator trains others in completing URM tasks. This may involve the preparation of training materials, scheduling training sessions, and providing extra help if someone does have difficulty with URM tasks.

Prepares Some URM Data. The coordinator is responsible for developing certain functional requirements data. In a way, the coordinator serves as a representative for requirements of shared spaces or spaces that belong to the entire organization, not to any one organizational unit.

Compiles Certain URM Data. Data prepared by other URM participants are submitted to the coordinator. The coordinator compiles the data into a final document or computer files, and may manage the work of clerks, computer specialists, artists, and others in preparing the final material.

Conducts Major URM Meetings. Besides training meetings, there are other times when URM participants will have to get together. Many of these meetings will be arranged for and chaired by the coordinator. These meetings may address interactions and relationships among various organiza-

(text continues on p. 35)

Figure 4-4. Sample memorandum designating a coordinator.

MEMORANDUM

To: All Department Heads and Staff Officers

From: Mr. A. J. Baker, President

Subject: Coordinator for Developing Requirements for Facility Project X.

1. The Board of Directors has determined that our company will move forward in planning for an expansion or replacement to our facilities. We will need to develop the functional requirements to establish in detail what we really need. Data will be prepared by every organizational unit in the company. We will be using a method called URM (user requirements method).

2. Mr. David Paul has been selected to serve as coordinator for the process of developing functional requirements. He will be reporting directly to me on this assignment.

3. Mr. Paul will instruct each organizational unit about what needs to be done, what staff assistance is needed, and the schedule of activities. The full cooperation of all organizational units is required.

4. We are very excited about the prospect of facility improvements. Our operations are currently limited by our facilities. I am sure everyone will give their full support to this effort and the leadership of Mr. Paul.

Figure 4-5. Major characteristics of a URM coordinator.

Leadership Qualities (1)*

– able to manage projects
– able to motivate others
– able to resolve conflicts
– respected by people in the organization

Continuity (1)

– likely to be around for the entire URM process
– available to represent resulting data to designers, top management, and others after data is compiled

Communication Skills (2)

– able to speak before groups
– able to conduct meetings efficiently
– able to conduct training
– know how to get messages sent within the organization
– able to prepare written memos, reports, and other material

Knowledge of Entire Organization (3)

– know organization from top to bottom
– acquainted with people in many organizational units
– general knowledge of operations and activities

Knowledge of Building Planning (3)

– general knowledge of building technology and construction
– general knowledge of building planning process
– general knowledge of budgeting process in organization

*Numbers in parentheses suggest importance; one is highest importance.

Figure 4-6. Major functions of a URM coordinator.

A. **Serves as the expert for the method.** Provides assistance to URM participants when difficulties with URM occur.

B. **Organizes implementation of URM.** Schedules steps, conducts meetings, and ensures that things are done on time.

C. **Trains all URM participants.** Trains representatives and others about their roles and responsibilities in URM.

D. **Compiles URM data.** Collects user requirements and supporting data from each organizational unit. Completes final documentation of user requirements. Handles the preparation of user requirements for all shared spaces or spaces not belonging to a particular organizational unit.

E. **Conducts meetings.** Conducts meetings that involve general issues for a project and issues affecting several organizational units. Examples of such meetings and issues are formulating the building objective, deciding who should be in a building, defining overall relationships, and resolving conflicts.

F. **Serves as the staff liaison for URM information.** Keeps top management informed about progress of URM. Serves as the point of contact for decision makers and designers about user requirements and supporting data during applications of URM. Gets conflicts resolved at the appropriate organizational level.

tional units and their requirements. Some of these meetings will be devoted to defining special relationships among organizational units. Meetings may provide feedback to URM participants about the effectiveness of their collective work or to iron out questions about the data submitted by certain organizational units.

Serves As Staff Liaison for URM Data. Someone needs to serve as a communicator to top management, designers, and others about the data contained in the final document. The coordinator has the most complete knowledge about the contents and can answer most questions about them. If necessary, the coordinator can call on representatives to explain details. As already noted, the coordinator is also best suited for explaining to URM participants what is happening with their data and the project they contributed to after URM data is compiled. In general, the coordinator is the logical person to call upon to explain the content of the resulting document, the procedures used, and the decisions that were made about handling data.

To perform these duties, the URM coordinator will need some assistance, at least clerical help. If URM data is compiled on a computer, a programmer or other computer specialist may be needed. (See Appendix B for suggestions about using computers with URM.) The staff required will depend on the size of a building project or the type of URM application and how quickly URM must be completed. For projects of a limited scale (those having less than 20 organizational units), the coordinator may be able to handle this job along with other regular job duties. Large projects will make the coordinator's job a full-time one. Top management must make sure that this person has enough time and staff to perform the job well. It is better to have staff prepared to assist and then not use them fully than to scramble for help once URM is underway. Large projects should not require more than two to four people. If there is a very tight schedule for URM, additional staff may be needed for the short, concentrated effort.

Others Serving as URM Coordinator

Staff facility specialists can easily assume the role of coordinator for URM. Design professionals from outside an organization can also fill the coordinator's role quite well. However, when professionals from outside the organization run URM, someone from a high level within the organization should help with coordinator activities. This is particularly true when the organization is large. This internal liaison person understands how the organization works, how to communicate within the organization, and how to gain cooperation of participants in the organization. The liaison person will make the external coordinator's work go smoothly.

Representatives of Organizational Units

An organizational unit is any formal or recognized element of an organization. Since URM works from the bottom upward in an organization, data are collected from each recognized unit, beginning with units at the lowest end of the organization's structure. Interactions among units are dealt with later.

A representative is needed from each organizational unit to gather data, complete analyses, define and log user requirements, and represent the organizational unit at URM meetings (see Figure 4-7). Each representative performs URM activities for the unit represented. Usually, this does not involve a great deal of time, except for a limited period of a few days or weeks. A representative can usually mix URM tasks with other regular job duties.

An organizational unit usually contains 5 to 25 people, although this number will vary. A representative is needed from each unit, regardless of where in the organization the unit is. For example, an organization may consist of a division with three branches and four sections in each

Figure 4-7. Major tasks of representatives of organizational units.

Develop User Requirements Data

- receive instructions on URM procedures
- prepare requirements data for organizational unit represented
- submit data to coordinator as instructed
- project future changes for the unit represented or know where to get help in projecting change

Verify Requirements Data

- obtain input and/or feedback from members and head of organizational unit
- review applicable elements of final documentation

Contribute to Interface with Others

- represent organizational unit in defining relationships among organizational units
- submit demands for shared space to coordinator
- represent organizational unit at meetings for resolving conflicts in requirements across units

branch (see Figure 4-8). Normally, a representative is needed from each of the 12 sections, from each of the three branches, and from the division, 16 in all. The representative of a branch develops requirements for only those things belonging to the branch itself, not requirements of sections within it. The branch representative may review requirements of sections in the branch, but does not develop section requirements. Also, assets used by all sections in a branch and belonging to no particular section are the responsibility of the branch representative. Similarly, the division representative deals with those things unique to or belonging to the division and not covered by individual branches.

Representatives of organizational units are normally selected by the heads of the respective units. Often the head of an organizational unit serves as—or at least selects—the representative. Characteristics of representatives are found in Figure 4-9.

Others Serving as Representatives

Staff facility specialists and design professionals can assume many duties of representatives, but a representative for each organizational unit is still needed as the source for information, unless a top-down approach is used. Instead of training representatives in compiling their own requirements, staff specialists and designers can identify and compile requirements and supporting data through one or more interviews with each representative. Specialists or designers must give a copy of any information derived through interviews to the representative so the data can be validated by the representative with others in the unit. Representatives are still the defenders and advocates for their organizational units and are probably needed at meetings dealing with overall relationships or resolving conflicts among units.

Figure 4-8. Example of an organization and its units.

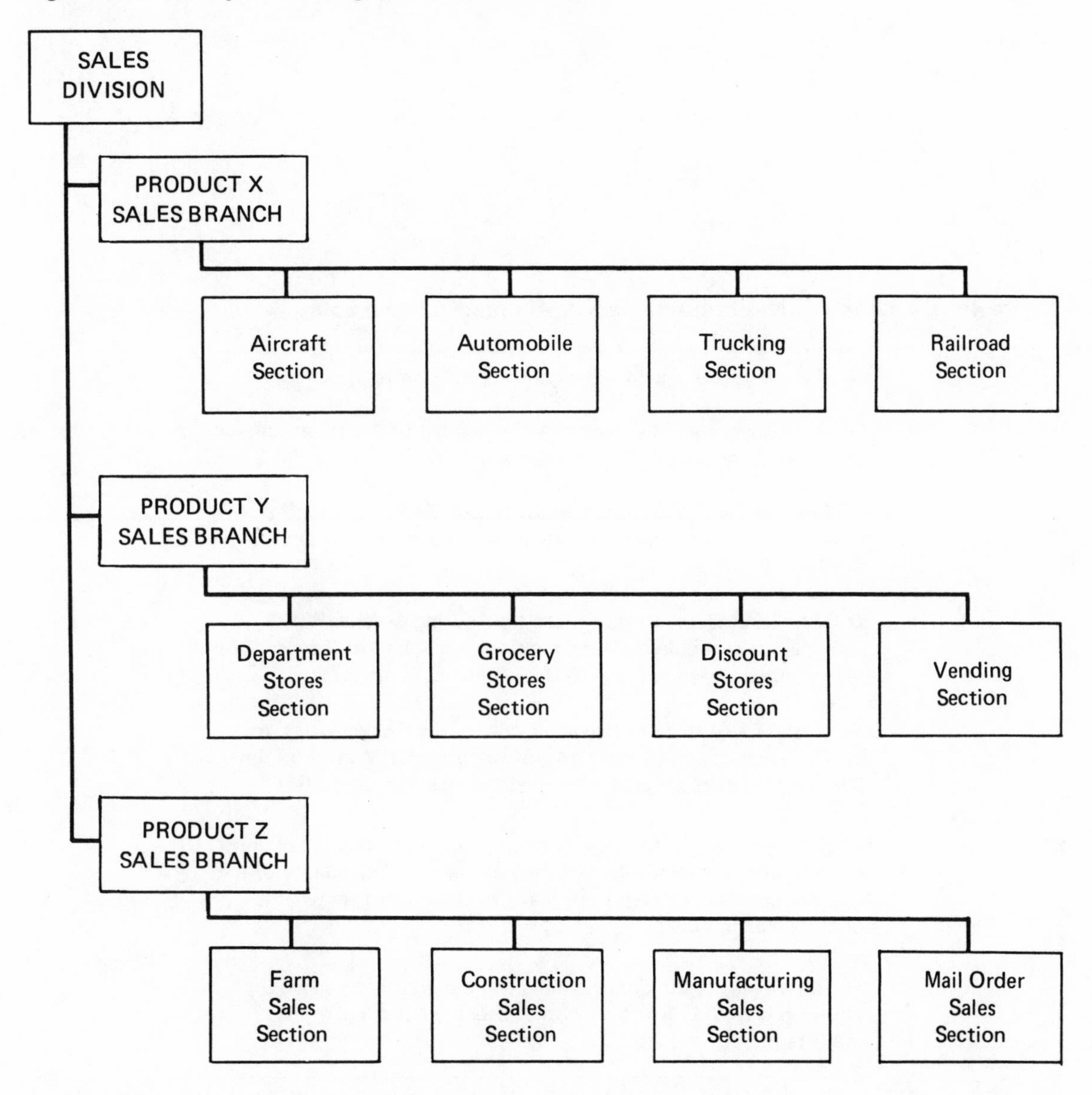

Figure 4-9. Characteristics of representatives of organizational units.

A. A good knowledge of the organizational unit represented.

B. A good working relationship with the head of the organizational unit (unless the head is the representative).

C. Get along well with others in the organizational unit. (This is important if the representative must collect information or get feedback from coworkers.)

D. Must be available for the entire time during which URM activities are scheduled. (Some continuity will be lost if a representative leaves during URM and must be replaced.)

E. Must be able to give sufficient priority to URM activities to provide adequate representation and complete URM tasks on time. (Only a few hours are usually needed each week during URM.)

F. Should be one of the capable people in the organizational unit, not one who is expendable or merely available. (The quality of the requirements developed is important for the unit to get what is needed in a building.)

G. Knowledge of building planning and design is not needed, although it would be helpful. Other characteristics are more important.

New processes and equipment may be proposed by designers and staff specialists as part of a facility project. Having designers and staff specialists serve as representatives of organizational units allows for new technology to be introduced in a coordinated manner across the entire organization or major portions of it. Certain organizational units may be eliminated or new ones created as a result of the process changes. Designers and staff specialists may want to introduce new processes and equipment to organizational units early in order to foster acceptance of changes. Coordinating changes with representatives will help identify detailed requirements that may otherwise be overlooked. Sometimes changes and effects on people are presented to organizational units after the overall solution is worked out.

Head of Organizational Unit

The heads of organizational units have two roles in the URM process. They must review information prepared for their respective units, and need to participate in resolving issues that affect their requirements.

Other People in Organizational Units

URM assumes that the users know the most about their activities and what is needed to perform them well. Very often members of an organizational unit know more about the details of an activity than the head or representative does. This knowledge should not be overlooked. Give members a chance to participate in the formulation of user requirements and at least review those prepared for their organizational unit. Members can contribute to accuracy and completeness of requirements.

Getting Organized for URM

With proper organization, URM can be implemented with few procedural problems. Organizing URM is primarily the job of the coordinator. Thus, selection of a coordinator is at the top of the list of actions for getting ready. Because the coordinator serves as the expert in URM, he should be well schooled in the user requirements method. A summary of steps in getting organized is found in Figure 4-10.

Define Objective

After a coordinator is selected, the first task is to draft an objective for the building. This should be done by top management, the URM coordinator, and by key technical personnel. While the final responsibility for a project rests with top management, the URM coordinator should be involved in defining the objective for the project because the coordinator will be the primary spokesman for the project for some time. Architects, engineers, corporate planners, or other key advisors may also be involved.

The main purpose for an objective is to clearly establish in written form a) why a building project (buy, lease, build, or modify) is being considered or needed, and b) why URM is being implemented. In most cases the need for a project is quite obvious and developing justification for the project or requirements to initiate a solution is the purpose for URM. In other cases URM is initiated so that the deficiencies of existing facilities can be precisely defined. The deficiencies provide the basis for making a decision to initiate a project and explore all possible solutions.

The objective is a brief statement explaining a) why a building is needed, and b) what it will be

Figure 4-10. Steps in getting URM organized.

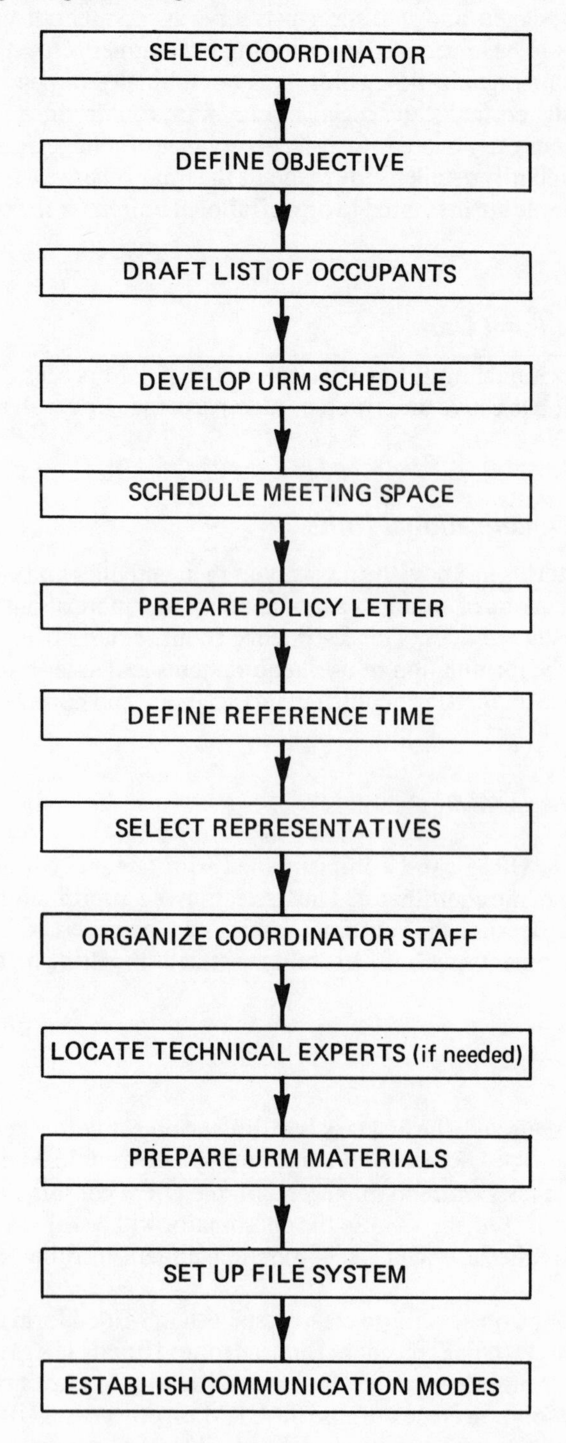

used for. At first, the group may find out that a draft statement is all that is possible because other analyses are necessary to clearly establish why a building is needed.

Two examples of objectives are provided in Figure 4-11.

There are many possible reasons for a building project. Some may have a technical basis, such as modifying or replacing an existing building to reduce high repair and maintenance costs or to meet current safety standards or energy conservation goals. An existing location may no longer be economical because of transportation, vandalism, depleted resources, or population shifts.

Reasons for a building project can also have a functional basis. An existing building is too small because the company or organization has grown. It may not have space for new kinds of equipment or activities. It may no longer be in a convenient location for citizens or employees. The production equipment may be inefficient.

The group defining the objective for the building must identify and at least express the major reasons for a project and what the building will be used for.

Draft a List of Occupants

Another early and essential piece of information about a building project is who will be in the building. At this point a final decision is not needed, but organizations or groups that are candidate occupants must be identified so they can be included in URM. The list may be analyzed later according to different combinations of occupants after requirements are compiled through URM. Then the list may be changed.

Normally, top management and the coordinator (possibly other advisors) will establish the initial list of occupants. The coordinator may recommend changes to it later.

Estimates about who will be in a building can be quite erroneous early in a project. For example, overlooking future occupants can cause major problems later. Problems with an existing building may be the center of attention. Out of those problems the need for a building project emerges. Attention is then given to a short-sighted solution involving only the organizations central to the problems of the old building. Initial estimates of project size and who will be in a new building are low. After careful analysis, the project grows to include many other groups. When analysts realize that there are important interrelationships among organizations not considered at first, there is a ripple effect.

Overestimates are also quite common. Everyone would like to be in a new building. As soon as the word gets out that a building project is imminent, managers at all levels begin to push for their own interests. It is considered unfair that some will have a new location and others will not. There is a lot of politicking and maneuvering to gain the best location. The individuals often lose sight of the good of the total organization. The easy solution is to include too many in the project.

A good estimate of future occupants is usually needed in order to apply URM. One reason is to get a representative on board for each organizational unit. Another reason is to discover the relationships among organizations that may exist and be an important factor in deciding which groups are to be in the new building.

In general, it is better to overestimate occupants early. The results of URM will provide an estimate of the size of the project and ultimately the funds required. Underestimating a project may require a second budget appeal or require some costly redesign or construction modification later on. However, it is more embarrassing to submit revised budget estimates because planners forgot to include some occupants than to submit a budget reduction due to original overestimates of needs.

Factors unique to each situation must be considered in deciding who should be included when URM is implemented. One organization was considering acquiring a new office building to relieve crowding and work environment problems. They had also started some renovation work on existing buildings. They chose to include all elements of the organization in URM and study the resulting

Figure 4-11. Examples of objectives.

OBJECTIVE

The objective for the municipal building of the Village of Tolono is to provide adequate permanent facilities for the Fire, Police, and Public Works Departments, the Village Board, President, and operating staff. The municipal building is a place where all elements of the local government are to be found and where most of the Village business is conducted. The municipal building should be a central focus (visual and operational) for the Village, portraying strength, symbolizing service and providing easy access to citizens.

OBJECTIVE

The objective for the application of URM for the XYZ Corporation headquarters is to define the administrative space requirements of all elements of the corporation located within 20 miles of the current headquarters. The resulting data will be used to determine: a) if a new headquarters facility is needed; and b) if so, what elements of the organization will be located in it, what elements will be located in existing buildings that will be renovated, and which of the current buildings should be torn down or sold.

requirements, and on that basis decided which elements would fill the renovated buildings. Requirements for the remaining elements would then form the basis for the new building.

When the preliminary list of occupant organizations has been made and after representatives from organizations are selected to participate in URM, it is a good idea to state clearly that a final decision about who will be in a completed project has not been made. It should be clear that the organizations represented are only candidate occupants.

Two examples of a list of occupants are found in Figure 4-12. The length of a list of occupants will vary with the size of the project. Included in the list is the number of personnel in each organization. The number of occupants in a building is a very important indicator of how large the project will be, but certainly not the only indicator. Data for other indicators are collected in subsequent tasks. The number of personnel may also be helpful in determining if more than one representative is needed to distribute work associated with URM. A rule of thumb is that a representative preparing data for more than about 20 to 30 people may not be fully knowledgable about the activities and needs of those represented. The coordinator will have to judge if additional representatives are needed for very large organizational units.

It may also be helpful to combine an organizational chart with the list of occupants to show how organizations are related administratively. There should be a clear correspondence between the organizations included in the list and those shown in the chart. The same level of detail should be used for both.

URM Schedule

Once the coordinator is selected and understands URM, further steps are taken to get ready for implementation. Plot the steps and tasks of URM on a calendar. The amount of time required for each step will depend on many factors. Due dates for budget requests or project milestones set by top management are most likely to drive the schedule. The size of the building project may have some effect, but not much, because work is distributed across representatives. Anywhere from two weeks to three or four months may be necessary to complete all steps. A sample schedule is found in Figure 4-13.

Directive

A policy or directive for implementing URM must be developed. It can be drafted by the URM coordinator, but must be signed, approved, or endorsed by top management. The policy must include a goal for the use of URM, a list of activities and functions, a schedule, assignment of responsibilities to participants, and a definition of their roles. It may also define funding sources for the URM process, space assignments for the URM coordinator and staff, the period for which the policy is in effect, and how participation in or performance in URM activities will be measured for heads of organizational units. A sample directive is found in Figure 4-14 (p. 46).

Selecting a Reference Time

URM is like taking a photograph of an organization. URM data represents an organization frozen in time. But organizations are continually changing. URM handles this by a) selecting a reference time for which requirements are developed, b) estimating what changes are likely to occur after that time, and c), if necessary, reviewing and updating requirements at a later date.

The important thing is to select the reference time before URM is started. In most cases one simply chooses the present. If it is known that a significant organizational or operational change will occur very soon, one can define requirements assuming the change has been implemented. In plan-

Figure 4-12. Examples of lists of occupants.

List of Occupants

Organizational unit	Number of personnel
Village Office	5
Village Staff	1
Police Department	3
Public Works Department	8
Fire Department	4
ESDA	2
Total	23

List of Occupants

Organizational unit	Number of personnel
Sales Division	5
Product X Sales Branch	3
Aircraft Section	6
Automotive Section	10
Trucking Section	12
Product Y Sales Branch	4
Department Store Section	15
Grocery Store Section	9
Drug Store Section	5
Vending Section	6
Product Z Sales Branch	3
Farm Sales Section	12
Building Supply Section	18
Manufacturing Supply Section	9
Mail Order Sales Section	24
Total	141

Figure 4-13. Sample schedule for URM.

Week 1. Select Coordinator

Week 2. Organize URM Process

Week 3. Send out URM Schedule and Preliminary Instructions

Week 4. Meeting 1—Complete Step 1

Week 5. Meeting 2—Complete Step 2

Week 6. Meeting 3—Complete Step 3

Week 7. Complete Documentation

Week 8. Present Results

NOTE: Actual schedules will vary considerably; activities need to be adjusted to fit schedule and the knowledge and experience of participants.

ning for URM, coordination with corporate or high-level staff is essential to select the correct time for the URM "picture" of the organization.

Sometimes one cannot establish which of two options for operational methods or organizational structures will be implemented. It may be best to develop two sets of requirements, one for each option. The results may help establish which option is better. Completing both sets of requirements will not delay the building project.

Participants and Training

The coordinator must identify each organizational unit that will possibly be involved and find out who the representative is to be. A memo can be sent requiring each head of an organizational unit to identify the representative for the unit. A sample is found in Figure 4-15. Create and maintain a list or directory of participants (see Figure 4-16). It is wise to include phone numbers and mailing addresses.

The coordinator should train all participants in how to perform URM activities. The reader will learn more about the URM process in Chapters 5, 6, and 7 or variations suitable for applications of URM in Chapter 8. One method is to train representatives in short sessions as URM is implemented. Figure 4-17 outlines a training program that is completed with five sessions distributed over the course of the URM process. Another approach is to have one long session during which all steps and tasks are covered at once. Figure 4-18 outlines a training program that can be completed with a full-day program. The training approach used will depend, in part, on how quickly the process must be completed.

At the beginning of training, have top management present URM and explain why it is important. This can serve to motivate the participants and help ensure their full cooperation.

Figure 4-14. Sample directive from top management.

MEMORANDUM

To: All Department, Branch, and Section Heads

From: A. J. Baker, President

Subject: Policy Regarding Participation in Developing Requirements for Project X

1. All organizational units will participate fully in developing requirements for our facility expansion project. Detailed instructions will be provided by Mr. David Paul, Project X Planning Coordinator.

2. Requirements development must be completed in the next four months. By the end of January all data needed for a final decision of the Board of Directors must be fully compiled and documented. Mr. Paul will provide you with a detailed schedule of activities. There will be no delays or extensions. Completion of data submittal will be added to the performance objectives of each Division, Branch, and Section supervisor.

3. Mr. Paul will need some supporting staff at the coordinator level. Department A will provide a clerk-typist for the four-month period. Department B will provide one programmer/computer specialist. Additional staff requirements may be placed on the Training Department, Art Department, Publications Department, and other units at a later date.

Figure 4-15. Sample memorandum for identifying representatives.

MEMORANDUM

To: All Division, Branch, and Section Heads

From: Mr. David Paul,
Coordinator for Facility Project X

Subject: Need for Representatives from
All Organizational Units

1. One of the first tasks for starting the planning process is to identify a representative from each organizational unit.

2. The head of each Division, Branch, and Section will select a person who will develop data, attend necessary meetings, and coordinate planning activities for your respective organizational unit. A representative from a branch will cover aspects not covered by representatives from sections within a branch. Similarly, a division representative will be responsible for aspects not covered by branches within it.

3. You will need quality individuals as your representatives. Some selection criteria are attached to help you. (Data from Figure 4-9 can be used as the attachment.)

4. Each Division, Branch, and Section head will submit the name of a respective representative to me, together with phone number, mail stop, and office symbol, not later than September 15.

Figure 4-16. Sample list of representatives.

Project X Representatives

Organizational Unit	Name of Representative	Phone No.	Mail Stop
Division A	Ralph Jones	X 1234	Bldg 1, B-43
Branch X	Gary Brown	X 5555	Bldg 2, F-20
Section X1	Mary Green	X 5051	Bldg 2, F-66
Section X2	Charlotte Ord	X 5934	Bldg 2, G-77
Section X3	Michelle Great	X 6611	Bldg 2, H-12

Meeting Space

A place is needed for training sessions and other meetings that will be required during URM. Suitable rooms should be available. Make arrangements for their use. During some meetings a blackboard will be needed. Other equipment, such as audio-visual equipment, may also be required. Secure them in advance to make sure the URM schedule is not hampered.

For training, a room with rows of tables is preferred so training participants have space for training materials. All should have a clear view of the front so that projected images and a blackboard can be easily seen.

When meetings for defining relationships among organizations are conducted, a U-shaped arrangement with a blackboard at the front is best. Participants need to see and interact with each other. The meeting leader will be at the front, sketching on the board until a suitable relationship diagram is completed. Further details about conducting these meetings are found in Chapter 6.

When meetings are used to discuss procedural problems, conflicts among organization unit requirements, or other matters, seating around a table is best. Participants facing each other are drawn into the discussion more easily.

URM Materials

During the URM process or URM training, participants will need to log data and perform analyses. This book contains examples of completed worksheets which are typical of those completed during URM. Special training materials may be prepared or one can purchase copies of this book to distribute to URM participants. Whether the blank worksheets found in Appendix A are used or others are created, copies must be made and distributed for use by participants. Other aids are also needed. These will include the requirements checklist (Table 6-1), the space estimating guidelines (see Chapter 6, Task 6), or similar materials prepared by the coordinator.

Figure 4-17. Multiple-session training program for URM.

Session 1. **Orientation** (Top Management)

- Attendance
- Background of Project
- Purpose of Effort
- Why URM is Important
- Introduction of Coordinator
- The Six Keys of Organizational Accomplishment (Coordinator)
- The PEAS of URM
- Overview of URM Process
- Responsibilities of Participants
- Schedule for URM Activities
- Where to Get Help

Session 2. **Preparing for Step 1** (Coordinator)

- Overview of Step 1
- How to Complete Forms
- Practice in Use of Forms
- Restate Due Dates

Session 3. **Preparing for Step 2** (Coordinator)

- Overview of Step 2
- How to Complete Forms
- Practice in Use of Forms
- Restate Due Dates

Session 4. **Preparing for Step 3** (Coordinator)

- Overview of Step 3
- How to Complete Forms
- How to Submit Documentation
- Practice in Use of Forms
- Restate Due Dates

Session 5. **Feedback on Completed Effort**

- Presentation by Coordinator
- Report by Top Management
- Awards/Recognition

Figure 4-18. Single-session training program for URM.

Morning

Introduction (Top Management)

Attendance
Background of Project
Purpose of Effort
Why URM is Important
Introduction of Coordinator

The Six Keys of Organizational Accomplishment (Coordinator)
The PEAS of URM
Overview of URM
Responsibilities of Participants
Schedule for URM Activities
Where to Get Help

Orientation to Step 1 (Coordinator)

Overview of Step 1
Walk through Step 1 and Related Forms in Detail with Examples

Break

Orientation to Step 2 (Coordinator)

Overview of Step 2
Walk through Step 2 and Related Forms in Detail with Examples

Orientation to Step 3 (Coordinator)

Overview of Step 3
Walk through Step 3 and Related Forms in Detail with Examples
Submittal Procedures

Afternoon

Practice Session (Coordinator)

Groups of two or three individuals try to go through the entire process, using each form. Choosing one organizational unit and one group office or work station as an example usually works best. The coordinator (and possibly a helper) works with groups as needed to answer questions.

Closing Segment (Coordinator)

Distribute any materials (extra forms, schedules, memos, etc.) needed during URM activities.

Review of URM schedule
Where to get help
Final questions

Files

User requirements and supporting data should be collected and stored in an organized manner. Obtain space in a file cabinet. Devise a system for managing collected data. Very often unforeseen changes in project plans, organizational structure, participants, and other factors require a return to original data and updating or modification. A good organization for and access to original data can save a lot of time later. A suggested structure for files is found in Figure 4-19. There are two major parts, URM and project procedural matters and resulting URM data. This general structure can be adjusted to meet the unique needs of URM projects and URM applications.

Appendix B discusses how computer data-base management systems can be used to compile, store, and retrieve URM data. Some may wish to compile portions of procedural matters on the computer as well. Data could include a list of representatives and other participants (see Figure 4-12), and a list of actions required and completed. Microcomputer spreadsheet programs are suitable for these. The need to use a computer for URM will vary with project size, computer availability, and personal preference.

Support Staff

In Step 3 of URM (see Chapter 7 for details) user requirements and supporting data are organized into a form convenient for designers, corporate planners, top management, and others. To prepare URM data in final form (document or computer files), identify and obtain clerical staff, computer specialists, and other needed skills. Define procedures by which representatives submit requirements and requirements data. Also establish how submittals will be processed by the coordinator and supporting staff. If the final product is a document, arrange for word processing or typing help, artwork, and reproduction.

Some representatives may not have adequate staff and skills and will depend on the coordinator and a central support staff to complete a document that looks uniform and is complete.

If a data-base management system is used to create computer files (see Appendix B for more details), central input of data will probably be best to ensure that data is entered consistently and correctly. Exact procedures will vary with software, computer equipment, and knowledge and skills required.

Work Space

To ensure that the coordinator and support staff can work as efficiently as possible and perform as a team, a suitable workspace should be provided. The larger the project and the shorter the time available to complete URM, the more important a good work environment, proper furniture, and necessary equipment become. This is particularly true during Step 3. Requirements will vary with each application of URM and the size of staff.

Technical Experts

URM may be applied to a wide variety of building types. For some, particularly those that are equipment-intensive, it may be necessary to have experts or specialists available to assist representatives in special aspects of URM data. For example, at the time a facility-improvement project is underway, office automation may be implemented or modified. A specialist in office automation systems or the vendor may have helped plan the new system or changes to an existing one. The specialist will have to help explain to representatives what they will be getting and the coordinator will have to explain how office automation data will be handled in URM.

Figure 4-19. A suggested file structure for URM materials.

Procedural Matters

A. General Project Information
B. Project and URM Schedule
C. URM Costs (requisitions, purchase orders, accounting data)
D. URM Participants (representatives, coordinator, staff, experts, others)
E. Correspondence
   (can be subdivided by organizational unit, type of correspondence, etc.)
F. Training Materials
G. Forms
H. Computer System (Design, operating procedures, etc.)

URM Data

A. Summary Data
   Tabulation of Total Space Needs
   Tabulation of Spaces Required by Organizational Units
   Overall Relationships Diagrams
   Tabulation of Future Changes and Impacts
   Other Tabulations as Needed
B. Background Data
   Objective for Building or Project
   List of Occupants and Numbers of People
   Overview of Uses, Functions, and Activities in Building
   (flow charts, organizational diagrams, process charts, etc.)
C. Organizational Unit Data
D. Shared Space Data
E. Other Data (such as special studies)

In a medical facility specialists may help plan such spaces as operating rooms or intensive care units. A safety professional, fire protection engineer, maintenance specialist, and other specialists may be needed to help users identify special requirements.

Establish procedures in advance for obtaining and using such help. This will minimize the demands on experts and their cost.

Communication

During the execution of URM, changes may occur in schedules or procedures and problems may arise. A method for communication among URM participants is particularly necessary for large organizations and complex building projects. Whether meetings are scheduled at regular times or whether company or electronic mail is used, keep communication channels open. Establish how information will be exchanged before URM is started.

Summary

Getting ready to apply URM is essential for making URM work smoothly. Participants need to understand the approach—the Six Keys of Organizational Accomplishment and the PEAS of URM. Staffing must be completed, whether an organization applies URM on its own or uses staff specialists or design professionals. Each participant, from top management to people in organizational units, must learn their responsibilities. A schedule for URM must be created. A policy statement or directive must be issued. A reference time must be selected. Training materials must be purchased or prepared. Details, such as a filing system, workspace, meeting space, and communication channels must be worked out. If needed, technical experts must be identified and brought in. When these matters are resolved, URM can move ahead smoothly.

Chapter 5

Analyzing Operations and Activities

In Step 1 of URM, the operations and activities which a proposed or current building are to support are analyzed. One cannot begin to state what is needed from a building until one clearly understands what will be in it. The analysis addresses the question, "How *should* operations be conducted and missions of occupants accomplished?"

Step 1 of URM is completed by having each representative perform the analysis in three tasks. In Task 1, each representative defines the mission of the organizational unit. The conceptual model of the Six Keys of Organizational Accomplishment (see Chapter 4) is used as a guide. Also in Task 1, the functions that are used to complete the mission are defined. In Task 2, representatives tabulate the PEAS for each function. In Task 3, representatives verify with others in the organizational unit that data are complete and correct.

Forms or worksheets are suggested as a means to assist in orderly data collection and analysis. Variations in forms may be required to meet unique application needs. Blank forms suitable for most applications are found in Appendix A.

Before representatives begin to use data forms, the coordinator must explain the reference time that will be assumed (see Chapter 4). Everything subsequent to the reference time will be handled as future data, primarily logged on Form H, discussed later in Chapter 6. Everything up to the reference time will be logged on forms which are explained below and in Chapters 6 and 7.

Record Missions and Functions (Task 1)

Activities in URM for organizational unit representatives usually begin with this task. If an organization is very small, representatives might have participated in getting ready for URM (see Chapter 4).

Mission Statement

The goal in this task is to create a statement that describes the mission of each organizational unit and to list the functions of each that will accomplish its mission. These data form the top of a hierarchically structured description of what an organizational unit is about (see Figure 5-1). Formulating these data causes organizational units to evaluate what they do and what they should be doing. Later, these data will establish what role a facility has for each organizational unit. Through self-evalu-

Figure 5-1. Hierarchy of terms to describe what an organization does.

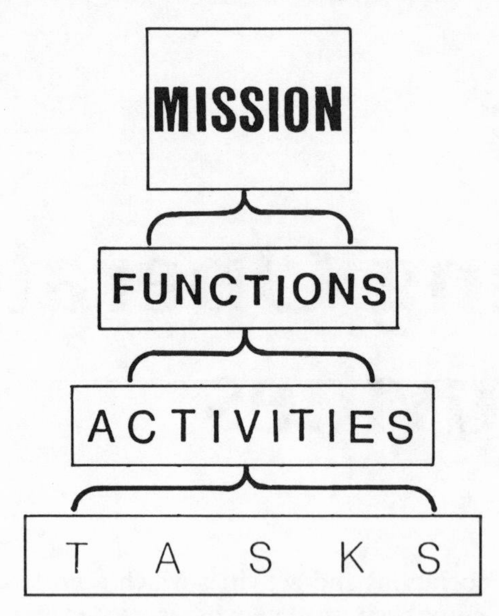

ation, representatives begin to discover whether they are actually doing what they should be doing before money is spent to provide a facility that supports ineffective activities and operations.

A *mission* is a statement that explains why an organizational unit exists. In many organizations, a formal statement of mission or purpose exists in a policy describing each organizational unit. A mission may be found in by-laws of a corporation or in the policy or document that first created an organization. Representatives may want to refer to such a formal statement of mission. However, the formal statement may not be up to date or fully reflect what an organizational unit is actually responsible for. In general, it is best to formulate a mission statement in laymen's terms, referring to formalized statements only for guidance.

The mission statement is logged on Form A together with other data, including the name of the organization, the name of the representative, and the number of people in the organization. Form A is formatted for this purpose. Examples are shown in Figures 5-2 and 5-3.

Remember that the application of URM is essentially the same for small and large organizations and facility projects. The method is applied primarily within each organizational unit. Examples illustrating the use of URM forms illustrate data within organizational units. Many examples could be given, but would differ only according to the function of the organizational units, not the size of the project or the size of the organization applying URM.

Tabulate Functions

The tabulation of the functions of an organizational unit is closely associated with the development of a mission statement. *Functions* are major classes of activities that are used to accomplish the mission. Functions are the second level of data in the descriptive hierarchy. For most organizational units, there is usually some administrative function. Other functions can be quite varied. Figures 5-2 and 5-3 also illustrate functions on Form A. Each function should be given an identifying letter (A, B, C, and so on) for convenience of later data tabulation and organization.

Figure 5-2. Example of a completed Form A for a fire department.

MISSION/FUNCTIONS

URM form **A**

organization
FIRE DEPARTMENT

representative
DAVID GREEN

personnel

male	10
female	2
total	12

mission

TO PROVIDE LIFE- AND PROPERTY-SAVING SERVICES TO THE COMMUNITY

functions

A. FIRE CALLS
B. RESCUE CALLS
C. ADMINISTRATION
D. EDUCATION AND PROMOTION
E. FIREMEN TRAINING
F. EQUIPMENT AND VEHICLE MAINTENANCE
G. STORAGE

subordinate organizational units

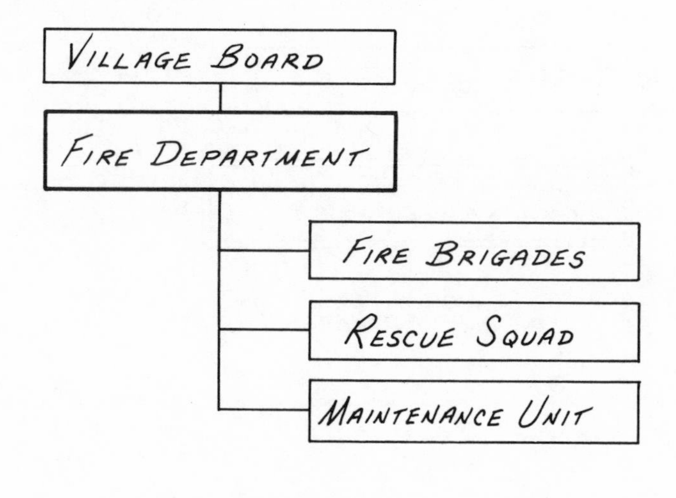

Figure 5-3. Example of a Completed Form A for a payroll branch.

MISSION/FUNCTIONS

URM form **A**

organization

PROGRAM BRANCH

representative

CHARLOTTE SHEPHERD

personnel

male	10
female	8
total	18

mission

COORDINATE THE DEVELOPMENT AND INTEGRATION OF COMPANY PLANS AND PROGRAMS RELATING TO RESOURCE REQUIREMENTS IN SUPPORT OF PRODUCT DEVELOPMENT AND MARKET TESTING.

functions

A. BRANCH ADMINISTRATION

B. PROGRAM ANALYSIS

C. PRODUCT DEVELOPMENT LIAISON

D. MARKET TESTING LIAISON

subordinate organizational units

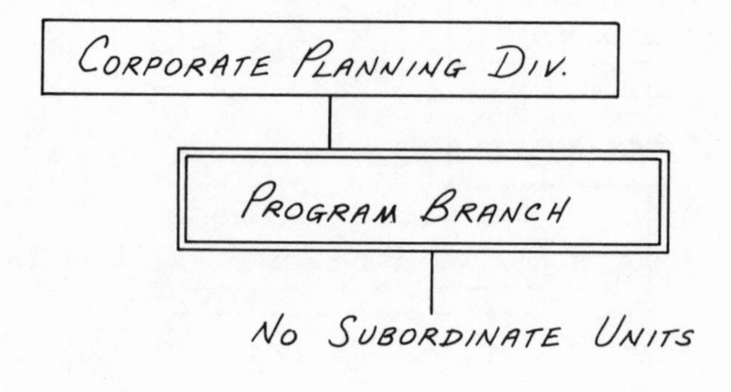

Segment of Organizational Chart

In order to help a designer (or someone else not fully familiar with each organizational unit) keep track of functional and operational relationships among organizations, it may be helpful to include a segment of an organization chart on Form A. Space is provided for this purpose. The segment should be limited to one level above and one level below (if they exist) the organizational unit described on the sheet. Formats can vary. Again, see Figures 5-2 and 5-3 for examples of a completed Form A.

If an organizational unit has no elements below it, an explanatory note can be inserted. If it is determined that functional relationships among organizational units are not important, that segment of the form can be left blank.

Define PEAS (Task 2)

In this task, the analysis of an organizational unit reaches the level of detail that will be important later for defining user requirements for a building. The PEAS (*p*eople or *p*ersonnel, *e*quipment, *a*ctivities, and *s*chedule) of the organization are defined and logged on Form B. It is best to use a different Form B for each function of an organizational unit.

Activities

First,* activities are listed, *Activities* are what are done to accomplish a function. Each activity should be numbered within a function (1, 2, 3, et cetera). This will aid in tracking data later.

One must make a judgment about how detailed activities should be. Anticipation of data associated with activities (people and equipment) will help suggest if one activity should actually be listed as two or two should be listed as one. The goal is to be clear for later evaluation of facility requirements. It is better to have a little too much detail at first than to be too general. Activities can be combined later if they appear to be too detailed.

Sometimes a function cannot be subdivided meaningfully into activities. The function is then repeated as the activity. When this occurs, the repeated function does not have to be numbered. The identifying letter assigned on Form A is sufficient.

People

After functions are defined, the number of people and the type of skill or job position associated with an activity or a group of activities are listed. If desired, the names of individuals can be used. Future positions and staff are not included here.

Equipment

Also tabulated on Form B are items of equipment associated with the activities. Equipment is most often brought into the building by users. This is not a list of everything one would like to have, rather a list of those equipment items which are essential or which already exist and will be brought along when people move in. Future equipment that will be arriving after the reference time identified by the coordinator is not logged here.

*Readers should not be confused by the order of the four elements inferred by the acronym PEAS. The order (activities, people, equipment, and schedule) seems to be the most logical when analyzing a mission or objective of an organization. As a result, one might use the acronym APES. The author chose PEAS as the more palatable memory aid.

In some cases new equipment will be purchased at the time a facility is constructed, such as new office furniture. Procurement is normally handled later and may be guided from data logged here. Equipment in production or warehouse facilities is often planned as part of the building. In these cases specialists usually define what is needed.

Because each situation for which URM is applied is different, the coordinator will have to instruct representatives regarding how equipment purchased with a project should be listed. In most cases, new equipment purchases will not be a concern in this task.

Not every little item needs to be listed on Form B. This is not a complete inventory. Items that are important include those that are quite sizable and those that depend on some system in the building so they can operate or operate safely. Normally, items which are built into the building are not included as equipment. They will be handled later with requirements. Table 5-1 lists some characteristics of equipment that are important for buildings. The list can be used to help decide if equipment items should be listed on Form B (see Figures 5-4 and 5-5 for completed examples of Form B).

Note that in examples of Form B, activities, people, and equipment are grouped into logical blocks. The horizontal blocks are created and separated by a strip of horizontal blank space extending across all three columns of Form B.

It may be difficult to log the number or quantity of some equipment items. One example is free-standing shelving. One can enter the number of full-height units of a given size. Another way is to list the amount of linear feet of shelving.

For reoccurring equipment items, standard names, codes, and units of measure should be established. This will help with the tabulation of items later. It may be helpful to create a special list of standard equipment items and corresponding data, using those names, units of measure, and codes. Table 5-2 provides an example of standard data for office furniture and equipment.

Table 5-1. Equipment characteristics which have an impact on building requirements.

A. OCCUPIES SPACE	(including maintenance space)
B. REQUIRES:	ventilation air conditioning sound control vibration isolation electricity (especially nonstandard voltages) water sewer gases (compressed air, breathing air, natural gas, oxygen, acetylene, etc.) fuel make-up air (for combustion, etc.) radio frequency shielding barrier guards access other
C. GENERATES:	heat noise vibration air contaminants (gases or particulates) solid waste other

Figure 5-4. Example of a completed Form B for a fire department.

People Equipment Activities Schedule — URM form B

organizational unit FIRE DEPARTMENT **function** MAINTENANCE

activities	personnel	equipment
1. SERVICE VEHICLES	1 MECHANIC	1 FLOOR JACK (2 TON) 1 TOOL CHEST CART
2. SERVICE EQUIPMENT	(SAME PERSON AS ABOVE)	1 PORTABLE WORKBENCH
3. CLEAN AND PREPARE HOSE FOR DRYING TOWER	(FIREMEN UPON RETURN FROM A FIRE)	3 PORTABLE HOSE REELS

comments

Figure 5-5. Example of a completed Form B for a payroll branch.

People Equipment Activities Schedule — **URM form B**

organizational unit *Payroll Branch* **function** *Timekeeping*

activities	personnel	equipment
1. Record Time	4 Data Clerks	4 Desks w/Chairs 2 Badge Readers 2 Medium Tables 2 Computer Terminals w/Stand 2 Visitor Chairs 1 Letter File
2. Validation	2 Clerk Analysts	2 Desks w/Chairs 6 Letter Files 2 Computer Terminals w/Stand 2 Personal Computers 2 Visitor Chairs 2 Telephones
3. Supervision	1 Time Records Supv.	1 Desk w/Chair 1 File Cabinet 1 Computer Terminal w/Stand 2 Visitor Chairs

comments

Table 5-2. Sample list of standard office furnishings that can be listed on Form B.

ITEM	*CODE*	*DIMENSIONS*
Desk—standard	D	34 × 60
Desk—executive	DE	40 × 78
Desk—single pedestal	DSP	34 × 45
Desk—typewriter	DT	34 × 60
Desk—with fixed return-left	DRL	68 × 60
Desk—with fixed return-right	DRR	68 × 60
Credenza	C	18 × 66
Table—credenza	TCR	18 × 66
Table—standard	T	34 × 60
Table—conference	TC	36 × 72
Table—small	TS	24 × 36
Table—medium	TM	34 × 45
Table—telephone	TP	18 × 24
Table—typewriter	TT	18 × 42
File—letter, 5-drawer	F	15 × 27
File—legal, 5-drawer	FL	18 × 27
File—letter, 2-drawer	F2	15 × 27
File—legal, 2-drawer	FL2	18 × 27
File—safe	FS	18 × 27
Chair—standard, desk	C	18 × 20
Chair—visitor	CV	18 × 18
Bookcase—3 sections	BC3	14 × 24
Bookcase—4 sections	BC4	14 × 24
Cabinet—storage (2 door)	CS	18 × 36
Cabinet—map	CM	42 × 54
Drawing Board (5-foot)	DB	40 × 60
Drawing Board (6-foot)	DB6	45 × 72

Schedule Comments Block

Time and scheduling data that are important for building use can also be entered as comments on Form B. Shifts, regular hours for activities, peak times for activities, and similar factors are important. For example, three people may perform the same activity, one for each shift, sharing the same work station. A footnote may explain this fact, so that later on, one understands that one, not three, spaces are required. Figure 5-6 illustrates this example. (Note the first asterisk.)

Another example is that of corridors and washrooms in schools, which see peak traffic and use between class periods and little during class times. A note about similar peak-activity periods would be important and should be entered in the comments block.

Special data or footnotes can also be entered in the comments section. References to equipment documents containing technical data that may provide important details for designers are included here. Figure 5-6 illustrates how to use the comments block for references.

The equipment listed on Form B could be extended to provide a complete inventory of equipment and furnishings. In addition, the comments space can be used to identify current location and future location and data necessary for scheduling a move. Normally, this much detail is not needed, except when the move is about to take place. URM data can be updated at the time that information for moving in is needed.

Figure 5-6. Example of the use of the "comments" block on Form B.

People Equipment Activities Schedule — URM form B

organizational unit DATA PROCESSING — **function** COMPUTER OPERATIONS

activities	personnel	equipment
1. OPERATE COMPUTER CENTER	3 OPERATORS*- LEVEL 3	1. HARRIS MODEL 78 COMPUTER**
		2. TAPE DRIVES
		3 DISK SYSTEMS
		2 HIGH-SPEED PRINTERS
		1 OPERATOR CONSOLE

comments * 1 PER SHIFT; THEY SHARE THE SAME WORK STATION
** TECHNICAL DATA (POWER, COMMUNICATIONS, SERVICE) ON ALL EQUIPMENT IN THE CENTER IS FOUND IN XYZ CORP. MANUAL C-21.

Summary

PEAS data are compiled on Form B. Several sheets may be needed for each function. Data should be grouped in horizontal blocks to indicate what people and equipment go with each activity or group of activities. A small space between these horizontal blocks will help keep data in order.

Verify Data (Task 3)

URM assumes that users are the best source of information about what they do. This assumption recognizes that experience and knowledge is distributed throughout an organizational unit. Each individual may have some information to contribute that others may not know. Representatives should let people check data they might know something about. Yet it could be very impractical to have everyone verify all data. If this is the case, a selective approach can be very helpful, giving forms with draft data to key people.

In many applications of URM, the coordinator will want to verify the objective for the building and the draft list of occupants with representatives. They may foresee difficulties that were not seen by others. Problems can then be spotted and resolved early. For example, who is included in the list of occupants may be contested by one or more representatives. Their complaint or question may be a valid one that was not fully considered by those who created the list. Raising and resolving such issues early in URM will reduce other problems later on. A sample memorandum for soliciting comments is provided in Figure 5-7.

Each representative should have the mission, functions, and PEAS data reviewed by individuals in each organizational unit represented. The representative may have overlooked important items or may have entered items incorrectly. These errors can then be quickly detected and corrected. Obviously, the head of the organizational unit will have final responsibility for the content of Forms A and B. A sample memorandum for requesting other information in an organizational unit to review data is found in Figure 5-8.

Heads of organizational units will want to review the data prepared by representatives of organizational units at least one level below them. Conflicts and redundancy in data will be recognized and can be resolved.

One kind of problem that can arise involves shared equipment or facilities. As a general rule, if equipment is shared by more than one organizational unit, it is best to list it as belonging to a unit at the next higher level in the organizational chart. For example, three branches within a division of an organization may all have a vital need for a certain computer. Each branch may feel that the computer belongs to its organizational unit. One group may program it, one may operate it, and one may maintain and service it. Functionally, the computer belongs to the division and not to any one of the branches. As a result, it should show up on the data sheets for the division and not on the sheets for any of the individual branches.

Other kinds of conflicts or operational problems will surface during reviews of data about the way things should be. For example, an organization might be logically reduced in size or function, or even eliminated, if it is properly structured and not constrained by present facilities. Processes and procedures, including paperwork, information, equipment, materials, or people flow might be improved if one does not have to work around current facility limitations.

Conclusion

There are three tasks in Step 1 of URM. They are aimed toward examining how each organizational unit should conduct its operations. In the first task, the mission of each unit is defined and the

Figure 5-7. Sample memorandum used by a coordinator to obtain feedback from representatives on the draft "List of Occupants."

MEMORANDUM

To: All Representatives Participating in URM

From: David Paul, Coordinator for Project X

Subject: Feedback on List of Occupants

1. Enclosed you will find a List of Occupants that has been prepared for Project X by the Executive Committee. The list is a draft, showing which organizational units should be located in the new facility at the new location.

2. You are to review this list and submit your comments to me not later than June 29. Please consider operational and functional impacts of the list as shown. Your comments should state a) what changes should be made and b) the reasons for the recommended changes.

Figure 5-8. Sample memorandum sent by a representative to coworkers seeking feedback on URM data.

MEMORANDUM

To: All Staff in the Laboratory Branch

From: Michelle M. Sweet,
Laboratory Branch Representative, Project X

Subject Comments on User Requirements Data Prepared for Branch

1. To date we have completed an analysis of our Branch's mission and functions, as well as the personnel, equipment and activities we will bring into the proposed new facility. I have recorded them in the standard format we are using.

2. Please review these data carefully. Note any missing information or errors in the data already recorded. Please explain the reasons for your comments. I will get back to you if I have any questions about your notes.

3. Your comments must reach me no later than June 3.

4. This is the first of several opportunities you will have to participate in defining the requirements for our Branch. Your help is appreciated. The more accurate our data are when we turn it in, the more likely we are to get the kinds of space and features that we need.

functions used to accomplish the mission are logged. Also, the operational relationships among units immediately above and below each unit are diagrammed. In the second task, the PEAS for each function of the unit are recorded. Finally, data is verified with other people in the organizational unit, particularly the head of the unit itself and the head of the unit above it.

The tasks involved seem simple enough, but in fact they may raise many difficult questions about the way things are and the way things should be. Changes across the entire organization or within certain elements may result. The need for new equipment and methods of operation may be recognized. Staffing adjustments and reorganizations may result from the analysis in Step 1. Other major adjustments may result before Step 2 is even begun.

Chapter 6

Defining User Requirements

The goal of Step 2 of URM is to identify and record user requirements based on the data developed in Step 1. In Step 2, the main questions to be answered are:

1. Where will the activities, people, and equipment be located? What spaces are going to be needed?
2. How should required spaces be related to each other?
3. What features or characteristics are expected for each space and the building in general?
4. What changes might occur in the future among activities, people, and equipment and what effects might these changes have on the spaces and their characteristics?

In order to answer these questions Step 2 is divided into 12 tasks, as noted in Figure 6-1. The needed spaces (rooms, areas, et cetera) are named in Task 1. The activities, people, and equipment that users will bring into the building and already identified in Step 1 are assigned to each space in Task 2. Relationships among spaces and organizational units are defined in Task 3. What the building must provide or control in support of equipment is identified in Task 4. Other user requirements are identified and logged for each space in Task 5. The size of each space is estimated in Task 6. If necessary, spaces are classified by type. Demands for spaces shared by several organizational units (such as conference rooms, eating facilities, and the like) are projected in Task 7. Future changes in activities, people, equipment, and time are projected in Task 8, and effects on user requirements are also identified. General relationships among organizational units are determined in Task 9. Conflicts within and among organizational units are resolved in Task 10. If necessary, special studies are conducted in Task 11 to identify requirements which cannot be determined otherwise. Finally, in Task 12, user requirements data are verified with people who are not directly involved in the URM activities.

Tasks 1 through 9 are completed in sequence. Tasks 10 through 12 may be interspersed as needed with other tasks in Step 2.

Again, to aid in the orderly development of user requirements, several forms or worksheets are suggested. Blank ones suitable for most applications of URM can be reproduced from Appendix A. If so desired, variations can be made to those illustrated.

Figure 6-1. The twelve tasks in Step 2 of URM.

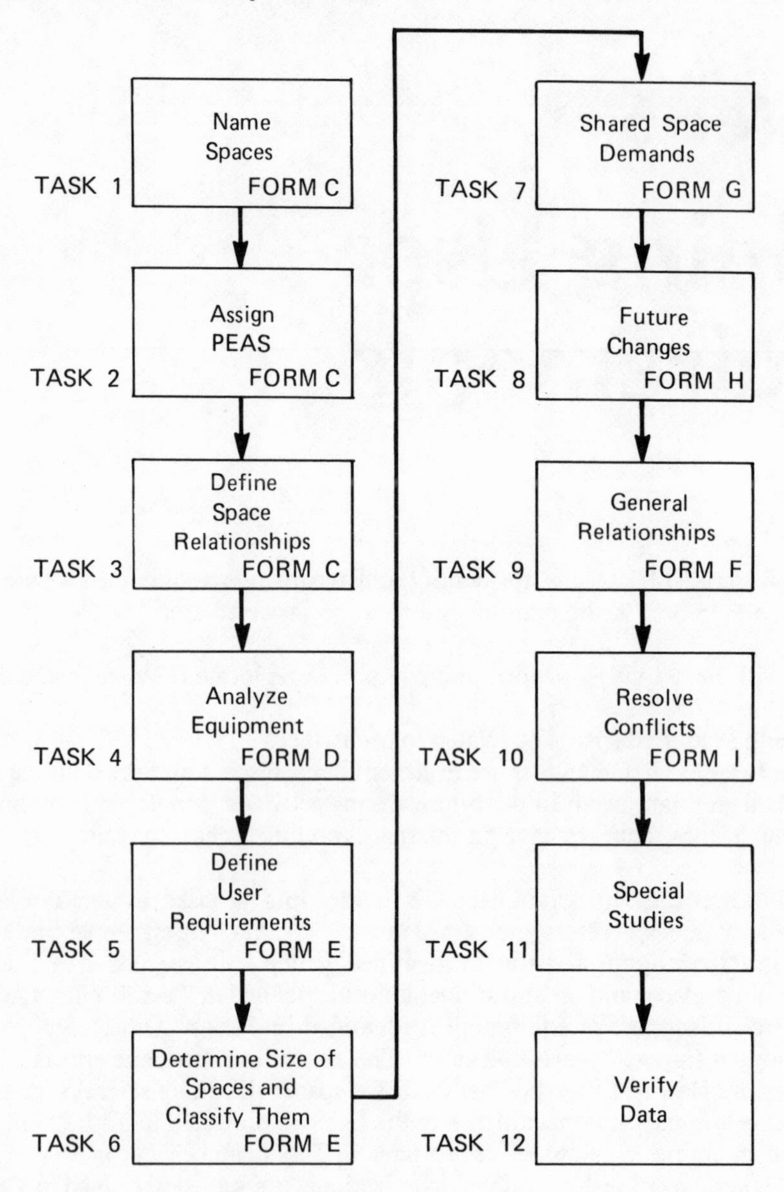

Identify Rooms or Spaces (Task 1)

Naming Spaces

In Step 1 representatives completed an analysis of PEAS (the activities, people, equipment, and schedule data) for their respective organizational units. In this task the question to be answered by each representative is "Where are the PEAS to be located?" Using Form C, the spaces required for each organizational unit are identified and tabulated. Only the "space name" column has to be filled during this task. Under "space name," Figure 6-2 illustrates how a Form C is filled in for Task 1.

Spaces are rooms, work areas, or locations associated with a building. Spaces can be indoors or outdoors. They can be bounded or unbounded, enclosed by walls or left open. Spaces can be thought of as floor area or volume. Each recognizable portion of a building is a space.

Spaces can be named in any manner meaningful to the people in the organizational unit. For example, an office might be the branch chief's office. There may be a tire shop and a welding area. There might be a storage closet or a computer work area. Any number of space types are possible. Outdoors there could be a receiving dock, a vehicle washstand, a parking lot, a lunch area, or a jogging track.

Remember, spaces do not have to have walls or be bounded. Some spaces may be distinguished only by the functions or activities that take place in them. Each representative must judge how detailed the list of spaces must be. For example, there does not have to be a separate space for each person in an open office area, particularly if people work as a unit.

Initially, all spaces required by an organizational unit can be listed on Form C. However, in Task 7, spaces shared by several units will be addressed. Some spaces may be deleted from the list on Form C if they are not the unique responsibility of and used very efficiently by the unit listing them.

Several units may assume that a space belongs to each of them. This kind of conflict (see also Task 10) can often be resolved in Task 1. It is best to attach a space that is used by several organizational units to the next higher unit in the structure of the entire organization. For example, a computer facility may be shared by three units—one responsible for developing programs, a second for implementations, and a third for maintaining programs and hardware. (In Chapter 5, this example was mentioned in the context of who "owns" the equipment. The same principles apply when considering who "owns" the space.) If one boss or administrative unit is over the three and has a representative in URM, the higher unit should list the space on its Form C, not the lower three units. The unit that is fully in charge of a space and its use is the one that should list it on Form C. If no one is fully in charge of a space and its use, then it is probably a shared space as defined in Task 7.

Numbering Spaces

On Form C, representatives give each space for their organizational units a name. For convenience in tracking spaces, each space can also be given an identifying number. Spaces could be numbered in different ways. One convenient way is for the coordinator to assign a unique one- or two-letter code to each organizational unit and have representatives assign a number to each space. The code for a space might then look something like "FD-3" or "L-11." The one or two letters refer to the organizational unit and the number refers to the third or eleventh space in the organizational unit's list of required spaces on Form C. An example of space identifying numbers added to the "Number" column on Form C is also provided in Figure 6-2.

Define Occupants and Contents (Task 2)

In this task the activities, people, and equipment identified in Step 1 are aligned with the spaces just named on Form C. The purpose is to note what belongs where. Another purpose is to see that all PEAS

Figure 6-2. Sample Form C.

SPACES/ADJACENCIES

URM form **C**

organizational unit: Fire Department

PEAS	number	space name
F1	FD-1	Maintenance Bay
F2	FD-2	Shop
F3	FD-3	Air Pack Charging Room
G	FD-4	Oil & Grease Storage
C1-3	FD-5	Administrative Office
A1, B1	FD-6	Dispatch Room
D, E1-2	FD-7	Training Room
E3	FD-8	Kitchen
A4	FD-9	Hose Drying Tower
H	FD-10	Building Entrance
A2	FD-11	Tanker Bay
A3	FD-12	Pumper Bay
B2	FD-13	Rescue Truck Bay

adjacency codes

0–same space
1–must be adjacent
2–close, adjacency preferred
3–anywhere nearby
4–distance unimportant
5–must be distant

are accounted for in terms of where they will be located. This task may require that some spaces be added to the list or deleted from it. An example of a Form C with codes listed in the "PEAS" column is also found in Figure 6-2.

Functions needed to accomplish a mission of an organizational unit were previously defined on Form A and activities which complete the functions were listed on Form B for each function. Each function was assigned an identifying letter and each activity for a function was assigned an identifying number. One or more activities and/or functions will be performed in each space just listed on Form C. The letter corresponding to each function and the number for each activity are entered in the column on Form C labeled "PEAS". No functions or activities should be left unaccounted for in matching data from Form B to spaces listed on Form C. Activities or functions must be located in some space. If functions or activities are left over on Form B, they either will not be in the facility and should be dropped from Form A (or an explanatory note added) or one or more spaces must be added on Form C for assigning the remaining functions and activities.

For example, Form A for the fire department listed "Equipment and Vehicle Maintenance" as Function F (see Figure 5-2). There were three activities listed on Form B for this function (refer to Figure 5-4). Each requires a different space: a maintenance bay, a shop, and an air-pack-charging room. The three spaces were listed on Form C (see Figure 6-2). The corresponding codes in the "PEAS" column on Form C were F1, F2, and F3 respectively.

If equipment or personnel associated with a function or activity are to be placed in two or more spaces, the functions and activities and the blocks of data associated with those activities need to be adjusted.

Equipment and supplies are usually kept in one space; they are not normally moved around. Conversely, people may move about and not have one regular location. People, job positions, or names of individuals should be listed with the space where they are most likely to be found. Later, on a different form, comments can be made about people who perform activities in more than one location or for equipment that must be kept in more than one location, so they are not listed twice. Duplicate records of PEAS should not occur.

Establish Adjacencies (Task 3)

The purpose of this task is to determine how spaces should be related physically. Only spaces within each organizational unit are considered at this time. Relationships among spaces belonging to different organizational units will be considered later. In working out a solution, a designer will arrange the spaces in floor-plan drawings, organizing them on one or more floors. Users must explain which spaces need to be next to other spaces so the designer can arrange them correctly.

The matrix on the right side of Form C is used to define relationships. In considering each pair of spaces, Form C suggests five kinds of relationships:

1. pair of spaces must be adjacent
2. adjacency is preferred, but it is not absolutely necessary
3. anywhere nearby is fine
4. distance is not important at all
5. spaces should be far apart or are assumed to be in different buildings

A sixth relationship (0, zero) is included on Form C so that if two spaces originally entered on the form should turn out to be the same space, one does not have to redo the form to make the correction. Merely entering a zero in the relationship matrix handles the problem.

A value is then selected for each of the pairs of spaces named on the form. The values are inserted in the matrix at the intersection formed by the diagonal rows extending from the two spaces.

Some people have a difficult time entering values in the matrix. It can be visually confusing. It is best to start with the first space on the list as a reference space. Compare each of the other spaces in the list to it. Then use the second space as a reference and compare each of the spaces listed below it to the reference space. Progress down the list, making each space in turn the reference space. Compare it to each space below it until the last pair has been considered. Figure 6-3 illustrates this process.

The process of filling in the matrix can be explained for the data shown in Figure 6-2. The space at the top of the list, the maintenance bay, is considered first. With the maintenance bay in mind, one considers the first space below it (shop) relative to the maintenance bay. It was known in this case that parts removed from a delivery vehicle frequently had to be carried into the shop to be worked on. Also, tools were stored in the shop and needed to be brought into the maintenance bay. As a result, the two spaces had to be next to each other. An Adjacency Code 1 ("must be adjacent") was selected as appropriate and entered on Form C at the intersection formed in the matrix by the diagonals extended from the two spaces.

Next the air-pack-charging room was evaluated relative to the maintenance bay. There were no strong functional relationships between the two rooms. It didn't make a lot of difference how the two were related spacially, except that the two were part of the maintenance activities that were usually performed by the same people. Therefore, "anywhere nearby," a Code 3, was selected as appropriate and entered on Form C.

The oil-and-grease-storage area was evaluated relative to the maintenance bay. Because the oils and greases were flammable fuels, they needed to be separated from other spaces. They were best thought to be located in a separate shed or lean-to outdoors. A "must be distant" (Code 5) was selected and entered on Form C.

Finally, the administrative office was evaluated. There were no functional relations between the two spaces in the Fire Department. Thus, a Code 4 ("distance unimportant") was selected and entered on Form C.

This completed the evaluation of spaces relative to the maintenance bay.

Spacial relationships for the shop were considered next. As a standard starting procedure, the space listed directly below it (air-pack-charging room) was evaluated first. (There is no need to look at spaces above the reference space. Either there are none or they have already been evaluated.) The only factor to consider was the fact that they are both part of the maintenance function. A Code 3 was selected and entered on Form C where diagonals from the two spaces intersect.

Similarly, the oil-and-grease-storage space was considered. A Code 5 applied again and was entered on Form C.

Finally, the administrative office was found to have the same spacial relationship for the shop as it had previously for the maintenance bay. A Code 4 was written in.

This completed relationships remaining for the shop.

Next, the relationships for the air-pack-charging room were studied. The first space listed below it was the oil-and-grease-storage space, for which a Code 5 applied again. Then a Code 4 was used for the relationship between the air-pack-charging room and the administrative office.

A fully completed Form C for the Fire Department is illustrated in Figure 6-4. As explained earlier, the PEAS codes use a letter to represent functions and a number to represent the activity within a function that will be located in a space.

Obviously, spacial relationships are not limited to the spaces within an organizational unit. Relationships will also exist among spaces of different organizational units. Relationships among organizational units are considered in Task 9.

Rating schemes other than the five-value scheme suggested on Form C can be used. A variety of them can be found in facility planning literature. Some use multiple ratings for each pair of spaces in a matrix in an attempt to make relationship data more precise.

At best, the relationship matrix is only a guide. All the factors affecting relationship values cannot

Figure 6-3. How to fill in a relationship matrix.

1. Relative to first space

space name
MAINTENANCE BAY
SHOP
1

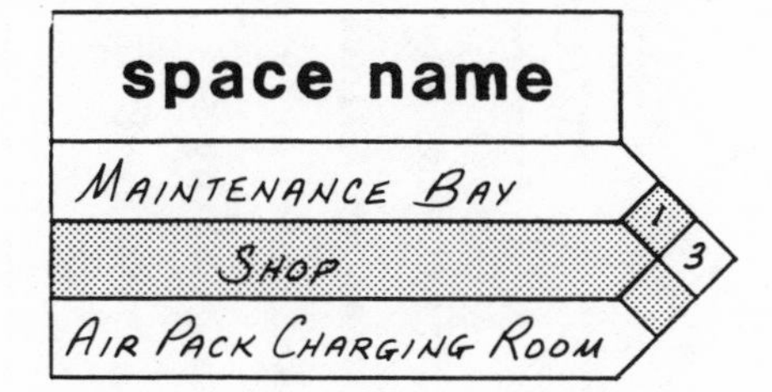

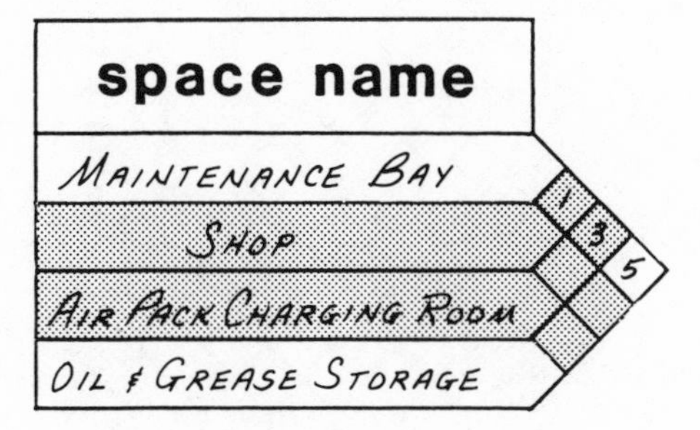

space name
MAINTENANCE BAY
SHOP
AIR PACK CHARGING ROOM
OIL & GREASE STORAGE
ADMINISTRATIVE OFFICE
1
3
5
4

2. Relative to second space

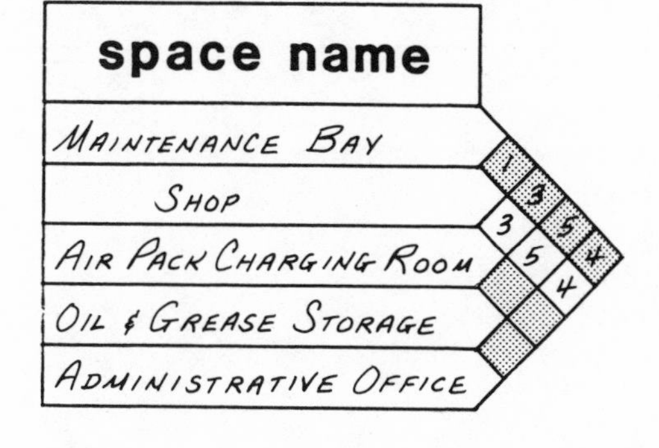

3. Relative to third space

space name
MAINTENANCE BAY
SHOP
AIR PACK CHARGING ROOM
OIL & GREASE STORAGE
ADMINISTRATIVE OFFICE
1
3
5
4
3
5
4
5
4

Figure 6-4. A fully completed sample Form C.

SPACES/ADJACENCIES

URM form C

organizational unit: FIRE DEPARTMENT

PEAS	number	space name	FD-2	FD-3	FD-4	FD-5	FD-6	FD-7	FD-8	FD-9	FD-10	FD-11	FD-12	FD-13
F-1	FD-1	MAINTENANCE BAY	1	3	5	4	4	3	3	2	3	2	2	2
F-2	FD-2	SHOP		3	5	4	4	3	3	3	4	3	3	3
F-3	FD-3	AIR PACK CHARGING ROOM			5	4	4	3	3	3	4	3	3	3
G	FD-4	OIL & GREASE STORAGE				5	5	5	5	5	5	5	5	5
C 1-3	FD-5	ADMINISTRATIVE OFFICE					1	2	3	4	1	4	4	4
A1, B1	FD-6	DISPATCH ROOM						3	3	4	1	4	4	4
D, E 1-2	FD-7	TRAINING ROOM							1	3	3	3	3	3
E3	FD-8	KITCHEN								4	3	3	3	3
A4	FD-9	HOSE DRYING TOWER									4	2	2	2
H	FD-10	BUILDING ENTRANCE										3	3	3
A2	FD-11	TANKER BAY											2	2
A3	FD-12	PUMPER BAY												2
B2	FD-13	RESCUE TRUCK BAY												

adjacency codes

0-same space
1-must be adjacent
2-close, adjacency preferred
3-anywhere nearby
4-distance unimportant
5-must be distant

be recorded and integrated mathematically or weighted systematically into one best solution. Whatever method is used, good human judgment is required on the parts of both the person entering values and the designer in working out a solution. In general, a designer will try to solve for all "adjacency required" relationships first and then move to each succeeding value. Regardless of what value scheme is used, care should be taken to ensure that the values entered in the matrix are reasonably accurate. A final determination of relationships will be made as design alternatives are evaluated and users state whether candidate designs and floor plans have acceptable arrangements among spaces.

Analyze Equipment (Task 4)

The purpose of this task is to establish whether equipment items will generate some condition that the building must control, or require something from the building in order to operate, or operate safely. Form D is used to help analyze equipment that will be brought into a building. One Form D is usually sufficient for each organizational unit. An example of a completed Form D is provided in Figure 6-5.

Conditions generated by equipment that may depend on the building for control include, but are not limited to, the level of noise, air contaminants, heat, moisture, and fire hazards. Sound control, ventilation and exhaust systems, air conditioning, and various sensor and alarm systems may be required.

Examples of conditions required for equipment include air conditioning, ventilation, lighting, electricity, water, compressed air, or special gases. Specifications for particular equipment items will define these requirements in detail.

Not all equipment and supplies listed previously on Form B in Step 1 need be included in this detailed analysis. Only those items that meet the two conditions noted (generate something that must be controlled by the building, or require something from the building to operate, or operate safely) should be evaluated. The idea is to alert the designer to potential demands on the building, not to develop solutions.

Sometimes items can be excluded from the analysis if conditions generated or required are considered routine. For example, most typewriters require 110 volts of electricity to operate. An assumption that all spaces will be supplied with 110-volt electricity can be made and recorded later in Task 10 to reduce the analysis of equipment required in this task and to simplify recording of requirements in Task 5. If such an assumption is made, all representatives must be informed so that the data are recorded in a consistent manner. Assumptions are written down and used in final documentation (refer to Figure 6-21).

If detailed data and specifications for equipment are known or reference data exist for it, a note can be entered or a reference document cited in the "explanation" column of Form D.

Identify User Requirements (Task 5)

General Procedures

In this task, each space and its PEAS logged on Form C are considered and the user requirements for the space are recorded. The question to be answered is "What must the building provide so that users can perform their functions and activities efficiently, economically, and safely?"

Requirements for major equipment were identified in Task 4 on Form D. Many other requirements exist to support performance of people and activities. In addition to performance, the satisfaction and preferences of occupants should be considered in defining requirements.

User requirements are logged on Form E. One sheet is used for each space listed previously on

Figure 6-5. Example of a completed Form D.

EQUIPMENT DATA

URM form **D**

organizational unit PUBLIC WORKS

equipment name	PEAS	conditions generated					conditions required					
		air contam	heat	noise	other	explanation	air cond	ventil	light	electr	other	explanation
AIR COMPRESSOR	B2			X							X	DISTRIBUTION LINES; TWO OR THREE OUTLETS
TABLE SAW	B1			X						X		220V
ELECTRIC WELDER	B2				X	ULTRAVIOLET RAYS				X		220V 3 PHASE
WATER METER WORK BENCH	C1								X	X		TASK LIGHT 110V FOR TEST DEVICES

Form C. Also in this task, requirements identified through the use of Form D are transferred to Form E.

To help in identifying requirements, a checklist of the most common user requirements (found in Table 6-1) can be used. The checklist does not contain all possible requirements. Others may need to be added for special situations.

Requirements may be logged as illustrated in Figure 6-6. It is helpful to group them logically by topic.

The requirements checklist (Table 6-1) was used to help define the requirements on Form E for the vehicle-maintenance shop. The first category thought important was access. A large overhead door was needed to get trucks and road-repair vehicles into the shop. To avoid having to open the large door for general access, a standard entry door was needed.

Several kinds of utilities were needed in the shop. Some 110-volt outlets were needed for power tools and drop cords and 220-volt, 80-amp service was needed for a welder. Water was needed to clean vehicles for service. Floor drains were needed to allow wash water or water from wet vehicles to drain away. Compressed-air lines and a compressor system were needed for work on tires, for cleaning parts, and for powering pneumatic tools.

At times engines would need to be run while vehicles were in the garage. Therefore, a system to capture and remove vehicle exhaust would be needed to ensure worker safety, one of the environmental requirements. Some heat would be needed in the winter to keep vehicles ready for emergency service. Another environmental requirement was good general lighting for repair work.

For communication, a phone was needed with an intercom capability to the office where job assignments were made and calls about field problems were received.

A few special features were also needed. Security of tool boxes (which belonged to individual service personnel) was important when workers left the shop. Eye-bolts secured to the wall were needed so that lock cables from tool boxes could be attached. Also, an overhead, half-ton hoist was needed so that engines could be pulled or other heavy lift work performed efficiently.

Other categories of requirements were judged not to be applicable to the shop.

Need and Purpose Codes

Although not absolutely necessary, codes can be attached to each requirement to indicate to a designer or someone who must respond to requirements whether they are negotiable and why they have been listed. Three need codes and six purpose codes are suggested on the bottom of Form E. The *need codes* tell a designer whether a requirement must be met or whether it can be deleted or modified if design problems arise. *Purpose codes* will help the designer understand why the requirements exist. More than one purpose code can be entered for each requirement. Figure 6-6 also illustrates need and purpose codes added to requirements.

Need and purpose codes were added to the requirements for the shop. For example, the overhead door was judged to be absolutely necessary (Code N), because large vehicles could not be moved into the shop if a smaller door were installed. The purpose for this requirement was to ensure that access was possible (Code 1).

For 110- and 220-volt electrical service, the need was absolutely necessary (Code N). Without proper electrical service, equipment could not be operated. The purpose for these requirements was to support equipment (Code 5).

Other Data on Form E

Blocks for four other data items are provided near the bottom of Form E. In Figure 6-6 these items are included in the example. Two of the blocks are provided for space-sizing data (size, such as square

Table 6-1. Checklist of common user requirements.

Requirements are statements of what is expected of a facility to support activities, equipment, or personnel. Requirements can be stated for an entire facility, major areas or space types, or for particular spaces or areas.

Requirements can be classified in many different ways; of course, categories always overlap to some extent. Requirements are varied for different spaces and activities. The list below is organized to help identify requirements which might be appropriate for particular spaces and their activities, equipment, and personnel. It doesn't contain all possible requirements; others may have to be considered to suit particular applications.

Space

Requirements in this category have to do with size, dimensions, and shape of spaces.

- Critical dimensions (height, width, length)
- Shape (rectangular, square, round)
- Clear span (minimum distance between columns or walls)

Access/Circulation

These requirements involve convenient movement of people or equipment within or into a space, control of such movement, or movement between spaces. Visual access (seeing in or out) is included.

- Privacy factors—sound control, visual control
- Size and type of openings (doors—width, height, type; windows—operable, customer, etc.)
- Control of openings (having a door, locks, etc. — related to privacy and security)
- Critical distances (horizontal, vertical) for cables, etc.
- Other functional relationships, such as access to a dock for forklifts, etc.

Utilities and Waste

These requirements are concerned with support systems which must be built into the facility. The type of system, the capacity or quantity to be handled by the system, limitations on its being shut down, tolerance or variance allowed, amounts of and locations for controls, and other performance characteristics for the system must be identified.

- Electrical service
- Water (hot or cold)
- Sanitary sewer
- Special sewer or waste system
- Solid waste system
- Special gases or fluids (compressed air, medical gases, etc.)

Environmental Conditions

Requirements in this category include conditions necessary for human occupancy, performance, and comfort, as well as for support of equipment. Again quantities, capacities, controls, limits, critical locations, and other data should be identified.

- Lighting—both general and task lighting
- Sound—control, levels
- Thermal conditions—heating and cooling, temperature, and cooling, temperature, humidity, air movement, and comfort ventilation
- Air quality—gases and particulates, dilution and exhaust ventilation, ventilation for hoods or booths
- Isolation and shielding from radiation, radio signals, etc.

Appearance/Finishes/Image

These requirements focus on the general character to be achieved (for image, safety, and morale of occupants), particularly for the surfaces.

- Characteristics or type of wall, floor, or ceiling (static free, washable, nonslip, color, wear and cleaning characteristics, loading or capacity —e.g., walls to accommodate charts and maps, or floors to accommodate forklifts)

Communication

These requirements identify built-in communication features or components for which supporting wiring or equipment must be included in design.

- Telephone terminals (detailed information is usually developed later in a telephone survey)
- "Hot" lines
- TV terminals or receptacles for monitors or cameras
- Microphone and speaker systems
- Computer networks

Storage

Built-in storage requirements can exist within a space or as a separate space. Requirements are easily handled in standard units of measure.

- Shelving or parts bins—total linear feet, height or number of tiers, special shelf depth
- Bulk storage—floor area required, height limits, special dimensions, cubic volume

Special Building Features

All built-in features not previously covered:

- Security features (vaults, safes, door locks, window bars, fireproof glass, special wall construction, heavy wire screens)
- Health and safety features which are built in (eye-wash fountains, emergency chemical showers, nonskid surfaces, barrier guards, etc.)
- Fire suppression or warning systems
- Lifts, cranes, hoists, elevators, ramps, docks, etc.
- Vibration isolation
- Other items important for user performance, satisfaction, or morale

Figure 6-6. A fully completed Form E.

REQUIREMENTS

URM form **E**

organizational unit PUBLIC WORKS

space name MAINTENANCE SHOP

requirements

ACCESS

- OVERHEAD DOOR (12'H x 12'W) N-1
- PERSONNEL ENTRANCE (3'W) N-1

UTILITIES

- 110V N-5
- 220V, 80 AMP N-5 (FOR WELDER)
- WATER (HOT & COLD) TO WASH VEHICLES PARTS, FLOORS N-1
- FLOOR DRAINS N-1
- COMPRESSED AIR (120 psi) I-1

ENVIRONMENTAL

- VEHICLE EXHAUST VENTILATION N-2
- HEATING (ABOVE FREEZING IN WINTER) N-5
- GENERAL LIGHTING N-1

requirements

COMMUNICATION

- PHONE (WITH INTERCOM TO PUBLIC WORKS OFFICE & VILLAGE OFFICE) L-3

SPECIAL FEATURES

- WALL ANCHORS FOR TOOL BOX LOCK SYSTEM I-4
- HOIST (1/2 TON) OVER 1 BAY I-1

Size 1280 SQ. FT. **Method** SKETCH **Space Type** SHOP **Class** 5

need codes
N-absolutely necessary
I-somewhat important
L-like to have, if possible

purpose codes
1-activity or function
2-health or safety
3-morale
4-security
5-equipment
6-other

feet, and the method used for making the size estimate). Procedures for making estimates of space size are discussed below in Task 6.

The remaining two blocks are labeled "space type" and "class." For some building projects it is important to classify all spaces named in URM. These two blocks can be used to enter classification terms or numerical codes. More than one classification for whatever purpose may be needed for each space.

On Form C, representatives named spaces in a manner meaningful to them. Any name could be used. Later it may be necessary to add up the area or volume required for all spaces of the same type. For example, totals may be needed later for all office space, all training space, or all laboratory space. Particularly if data is stored in the computer, lists of similar spaces, their size, and the total amount of space for the type of space can be created.

If a classification scheme is used, the coordinator should develop a list of space types or classification terms and a definition for each. The list should be distributed to all representatives, so that spaces can be classified or coded uniformly and accurately.

Repetitive Requirements

When there are many similar spaces, tabulating virtually the same requirements for each of these spaces will be time-consuming. The task can be simplified if standard requirements tables are developed and used. This process is described in Task 10. A sample list of standard requirements is found in Figure 6-20.

Space Layout

One kind of user requirement is the arrangement of items in a space. This is particularly important where arrangement is critical to the activities that take place in the space. Examples are customer and service counters. For most spaces a layout sketch is not needed. When arrangement is important, a sketch can be made to record ideas for layout. Form G can be used for this purpose. It should then be labeled as a "space layout" diagram in the space marked "Type of Diagram."

When making a layout diagram, avoid drawing rectangles. Use circles and ellipses for elements in the sketch. The purpose of the sketch is to record and communicate ideas for solutions, not to work out the solution precisely and in scale. Drawing rectangular forms introduces problems of scale and dimension for many people that do not occur when using circles and ovals. Scale and dimension are not important in these sketches. General ideas are their only purpose.

Begin with a large circle or ellipse to represent the room. Then use small circles or ellipses to represent furniture or equipment located in the room. A pair of lines intersecting the room boundary can identify where doors or windows should go if their presence and locations are important. Items can be labeled as needed to clarify the components of the sketch. Examples of layout sketches are found in Figures 6-7 and 6-8.

Figure 6-8 shows a layout sketch for a parts department room. The organization of elements in the room were thought to be important enough that a sketch was needed. Keeping customers out of the work and storage areas was very important. Furniture and paneling was suggested to create the barrier. Placement of displays visible to customers was also important. Separate customer and service entrances were also needed. These and other related elements were placed in the sketch and labeled. Initially, a circle was drawn to form the room. Then oval and round shapes were used to give approximate locations for the other major elements. Parallel lines intersecting the circle were placed and labeled to explain approximate door locations. Little attention was given to getting things drawn to scale, since that was not necessary to convey fundamental layout concepts.

Figure 6-7. Sample space layout sketch on Form F for a parts department room.

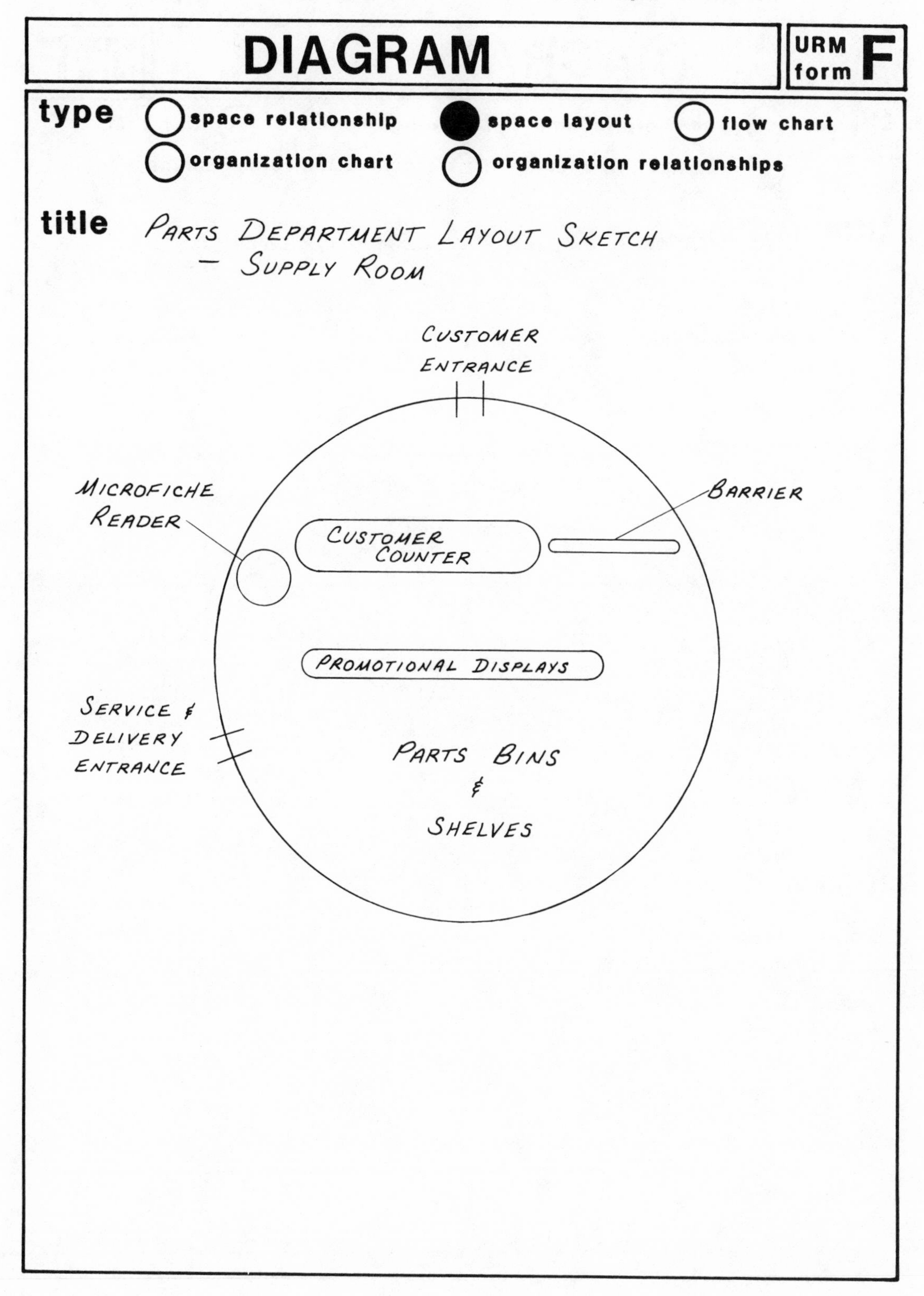

Figure 6-8. Sample space layout sketch on Form F for a fire department's dispatch room.

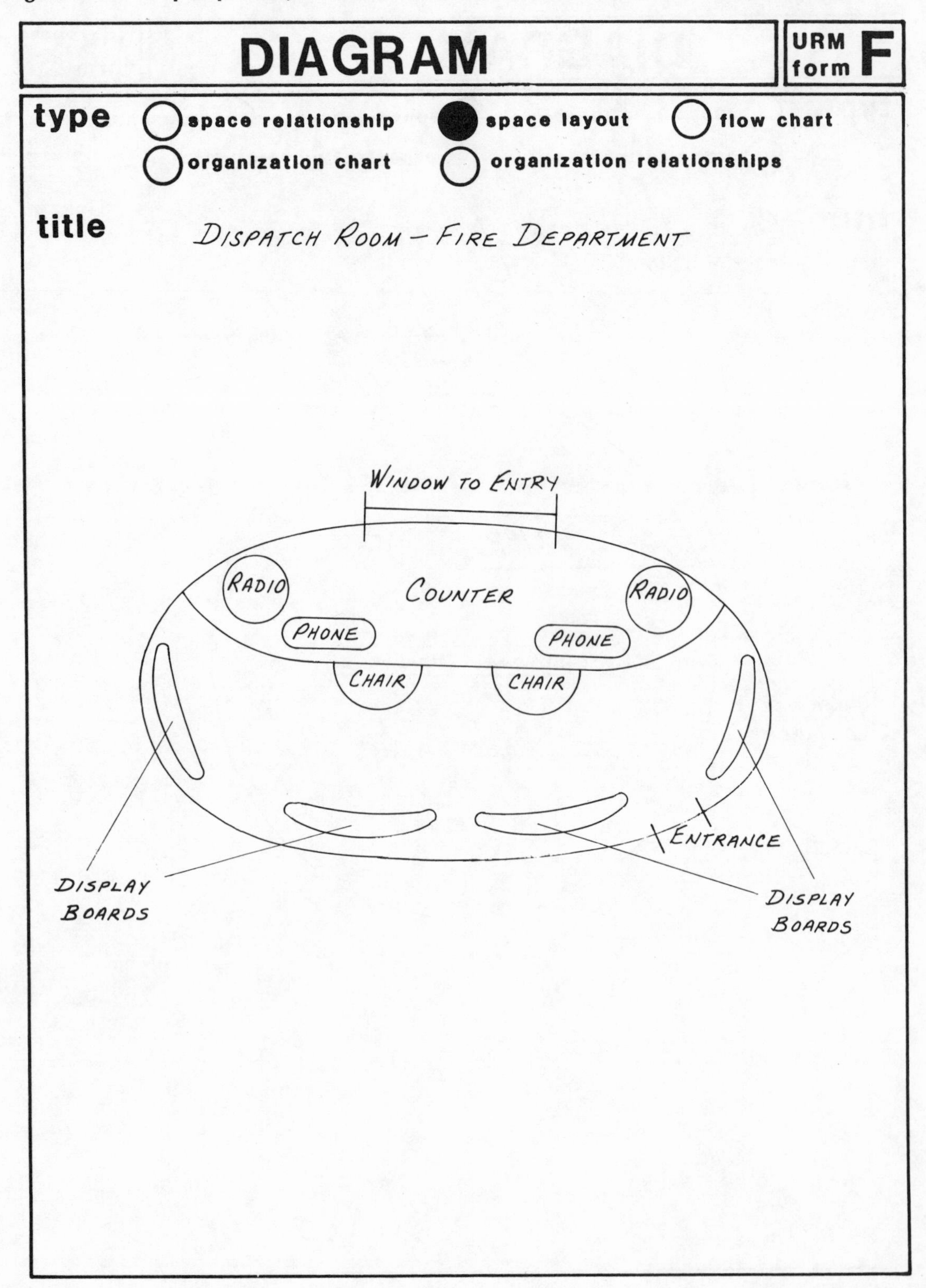

DIAGRAM URM form **F**

type ○ space relationship ● space layout ○ flow chart
○ organization chart ○ organization relationships

title DISPATCH ROOM – FIRE DEPARTMENT

In Figure 6-8 a dispatch room for the fire department is illustrated. Volunteer fire fighters enter the station and proceed to the dispatch room to find out where the fire is located and what equipment should be used. Fire locations are marked on large display boards showing maps of the area served. One or two dispatchers sit in the room, receive calls by both phone and radio from fire trucks, update the displays, and instruct arriving fire fighters. The sketch illustrates how equipment and elements in the room must be organized for effective use.

Estimate Sizes of Spaces (Task 6)

The objective for this task is to estimate the size of each space. Like other tasks in this step, Task 6 is completed by representatives. This is one of the more difficult tasks in Step 2. Representatives should seek help from the coordinator or people in their organizational unit if they have difficulty. The coordinator should be sensitive to the potential problems representatives may have and be ready to provide or obtain assistance for them. If available, experienced designers and staff specialists are the best source of help.

For most kinds of spaces, the floor area is the unit of measure. For warehouse and storage space, volume (that is, cubic feet) is a common measure. Developing an accurate estimate is quite important because, once finalized, estimates may not be reevaluated. In addition, after estimates are totaled to determine the size of a project, space is converted to project cost. If there are sizable errors in space estimates, significant errors in cost estimates will result.

A number of methods are available for estimating space size. Four methods are described in detail below: the space-standards method, the sketch method, the analytical method, and the comparison method. Others are possible. Some methods are easier to apply than others. Some are suitable only for certain kinds of space. These methods also vary in accuracy.

A size estimate is made for each of the spaces required by an organizational unit. The size and unit of measure are entered on each Form E. It may be wise to keep computation sheets until data have been verified (see Task 12). It may also be helpful to list on Form E what method was used to determine the size of a space. Refer to Figure 6-6 to see where the data is entered.

Space Standards Method

For many kinds of spaces, standards exist to estimate how much space is needed or allowed. These standards are found in architecture and design books, publications about particular building types (that is schools, offices, restaurants, et cetera), and other sources. Some companies and many government organizations have established standards for many kinds of spaces.

In the space standards method, the number of occupants or space units is multiplied by the standard to determine the amount of space required. An example for a theater or auditorium is found in Figure 6-9.

For the theater auditorium in Figure 6-9, the amount of space occupied by each patron is the same. If one knows how much space to allow for each patron, it is easy to multiply the space per patron by the number of people the room should hold to get the total area for the entire seating area of the theater. This assumes that the space estimate per person includes a portion that can be devoted to circulation or aisle space. Looking in architectural design-standards books, one can usually find space standards like those needed for theater seating. Here 12 square feet per person was used together with a capacity of 200 people, yielding 2,400 square feet.

The stage area is not included in the space standards for seating. Based on similar auditoriums, it was determined that for the kind of use this room would have, a shallow stage (10 feet) was adequate. A

Figure 6-9. Example of the use of the space standards method for an auditorium.

Auditorium

Number of Occupants: 200

Space Standard: 12 sq. ft. per person

Seating Space Required: 12 sq. ft. per person X 200 seats = 2400 sq. ft.

Stage Space Required: 10 ft. X 40 ft. = 400 sq. ft.

Total 2800 sq. ft.

Figure 6-10. Example of the use of the space standards method for an executive office and a classroom.

EXECUTIVE OFFICE

Rank: Division Head

Number of Division Heads: 4

Space Standard: 250 sq. ft. per division head

Space Required for Division Heads: 250 sq. ft. X 4 division heads

= 1000 sq. ft.

CLASSROOM

Type of Instruction: Lecture

Class Size: 40 students

Space Standard: 25 sq. ft. per student

Room Size Required: 40 students X 25 sq. ft. per student = 1000 sq. ft.

40-foot-wide by 10-foot-deep stage area was used, a 400-square-foot area. The entire room was then 2,400 plus 400, or 2,800 square feet.

Space standards that list the area required or allowed for particular kinds of rooms or spaces should be applied carefully. They are based on assumptions that may not be applicable to the space at hand. If possible, check to ensure that activities, equipment, and people match those assumed for the standards. If the assumptions for a space in a reference table match those of an organizational unit needing a similar space, the standard size can be used.

Some standards are upper limits, the maximum allowed. Care must be taken in applying them as well, because the upper limit may be more or less than that needed. Legitimate variances may need to be pursued, if activities, equipment, and people to be located will not fit in a space or if performance is significantly affected. Other space-estimating methods may be applied to see if a standard is reasonable.

Examples of the space-standards method for an executive office and a classroom are found in Figure 6-10.

For the executive office a company had a standard that was an upper limit based on rank. Division heads were allowed 250 square feet for an office. Thus, four division heads would be authorized 1,000 square feet.

The classroom is similar to the theater. Typical class size was assumed to be at or near 40 students. For lecture seating, standards suggested 25 square feet per student, including circulation and instructor space. Thus, 40 students times 25 square feet per student yielded an estimate of 1,000 square feet.

Sketch Method

In this method a scaled layout of a space is made. Contents are plotted first on grid paper at some selected scale. After contents are arranged in a satisfactory way, a rectangular box is drawn around the items. The scaled length and width of the box is determined by measurement. Then the length is multiplied by the width to establish the area required for the space. An example of this method is illustrated in Figure 6-11.

For this example of a computer room, the size of each machine was established by measurement or from the manufacturer's literature. An allowance was made for opening maintenance panels and using the machines. Templates were made of the use space with actual machine dimensions drawn in. The use space could be overlapped with adjacent machines, but machines could not extend into the use space of other machines. After a reasonable arrangement was arrived at (this does not have to be the exact, final arrangement) for all items of furniture and equipment, a box was drawn around the outer edge of the templates. The scaled dimensions of the box were determined to be 12 feet and 18 feet. Their product yielded a space estimate of 216 square feet.

In using this method, make sure adequate area is provided to use equipment and perform activities. Enough distance must be allowed between items for traffic and circulation. Access space for servicing equipment and opening doors is also important.

The main difficulty with this method is that not everyone can do a scaled layout. One must be able to visualize the way things should be in a plan view. Accurate measurements for equipment are needed to prepare a scaled sketch. Experience in doing layouts will help one determine how much space is required for circulation and traffic. However, since this is only an estimating method, it is probably better to be a bit generous than skimpy for traffic and other activities.

In using the sketch method, some people like to make scaled cutouts of items to be placed in a room. They can be moved around on grid paper until a suitable arrangement is found. They can then be drawn on the paper and dimensions for the space can be measured. Drafting templates with standard furniture measurements are available for making this kind of scaled layout.

Figure 6-11. Example of the use of the sketch method for estimating space size.

To apply the sketch method, the following steps are suggested:

1. Using grid paper (1/4 inch = 1 foot or 1/8 inch = 1 foot are commonly used for building layouts), establish an approximate layout for equipment. (Scaled shapes of equipment items can be cut out and placed on the grid paper. These can be moved around until an acceptable layout is achieved. Then their positions can be recorded on the grid paper.)

2. Determine by measurement or the use of scaled shapes whether activities can be performed in the layout—items moved around, equipment, etc.

3. Adjust the layout until a satisfactory arrangement is achieved.

4. Define the boundaries of the space by drawing a line around the layout (a rectangle is best). (Location of doors and other design features are not important for determining the size of the space.)

5. Compute the amount of space required by measuring the two sides of the rectangle to determine the length they represent in feet and multiply the two dimensions of the rectangle to find the amount of space required in square feet.

Note: The sketch resulting from this sizing method has no value except for deriving a size estimate. One does not have to include building features (doors, windows, etc.) unless they will definitely affect the size of the space.

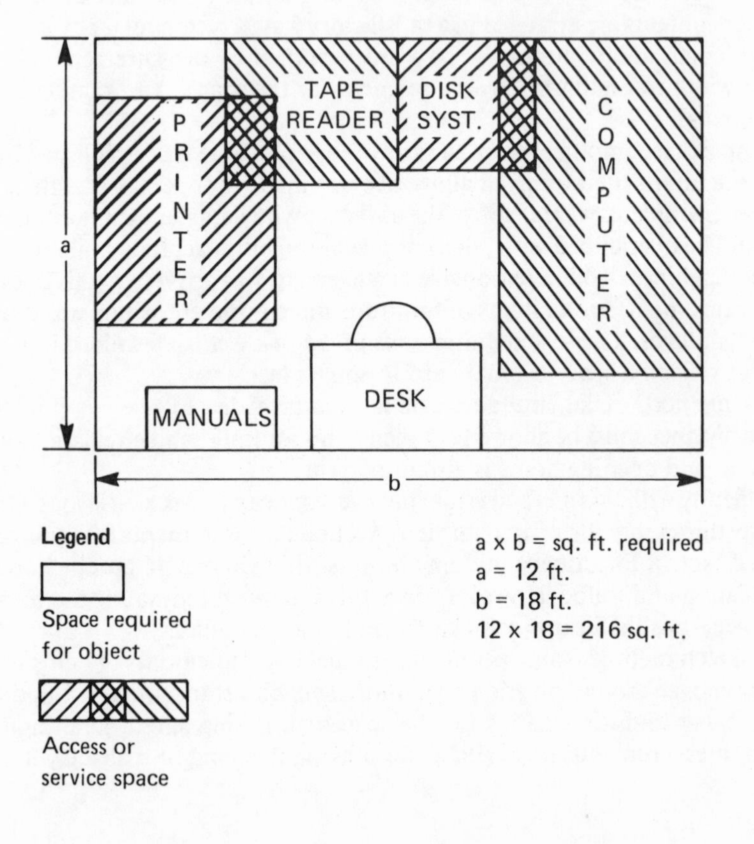

Analytical Method

This method is suitable for equipment-intensive spaces. A list of kinds of equipment and furniture is made. This list is the first column in a table used for analysis. The space occupied by each item is determined and included in the table. Then the amount of space required to use the equipment or furniture items (called *activity space*) is estimated and added to the table. The number of each kind of item is multiplied by the space needed per item. The space requirements for each kind of item are summed up to get a total demand. Finally, a circulation factor is added to complete the space estimate. Two examples of this method are found in Figures 6-12 and 6-13.

In the first example, the size of a two-person office space is estimated. It will contain three kinds of furniture: a standard desk (3 feet by 5 feet), a standard file cabinet (25 inches deep by 18 inches wide), and a visitor chair (about 2 feet per side). Areas were 15, 3.75, and 4 square feet respectively. To the dimensions of each of these is added an activity space on one side to access them (15, 4.5, and 4 square feet respectively). The total amount of space was determined by adding the area occupied by the furniture to its activity space, then multiplying by the number of each item (2, 3, and 2 respectively) and totaling the resulting space for all three kinds of furniture (100.75 square feet). An adjustment for circulation was added (40 square feet per person). This yielded a total of 180.75 square feet for the room. The .75 square foot could be dropped to avoid dealing with decimals.

For the shop there were again three kinds of equipment: drill press, lathe, and work bench. The size of each was determined and entered in the table. Considering the kinds of materials that would be

Figure 6-12. Example of the use of the analytical method for estimating space size (for a shop).

Item	(e) Floor area required (sq. ft.)	(a) Activity space required (sq. ft.)	(e + a) Total per item	Number required	Total space (sq. ft.)
DRILL PRESS	3x4=12	3x6=18	30	1	30
LATHE	2.5x6=15	(6.5x9)-15=42.5	57.5	1	57.5
WORK BENCH	3.5x8=28	3x8=24	52	2	104
				Total	191.5
				Circulation Factor	50%=95.75
				Grand Total	287.25

Figure 6-13. Example of the use of the analytical method for estimating space size (for an office).

Item	(e) Floor area required (sq. ft.)	(a) Activity space required (sq. ft.)	(e + a) Total per item	Number required	Total space (sq. ft.)
DESK	3x5=15	3x5=15	30	2	60
FILE	25/12 x 1.5 = 3.75	3x1.5=4.5	8.25	3	24.75
VISITOR CHAIR	2x2=4	2x2=4	8	2	10
				Total	100.75
				Circulation Factor	40 SQ/FT/PERS X2 = 80
				Grand Total	180.75

processed, an activity space was estimated and added to the table. Floor and activity areas were added and multiplied by the number of equipment items. This yielded a total of 191.5 square feet. Because some large pieces of material would be handled in the shop, a fairly generous circulation area was added (50 percent). A total of 287.25 square feet was estimated for the room. The .25 square foot could be dropped from the estimate for convenience or the number rounded off to 285 or 290 square feet.

Comparison Method

The comparison method is most suited to spaces which already exist and have contents as they will be in a new location. In the comparison method, the current space is measured and its size is used as an estimate for a required space. If a space being sized does not currently exist, or does exist and will have a number of changes in furniture, equipment, or activities, then one of the other methods should be used.

The main error in using this method is not making a careful comparison of some existing space to that required. It is an easy way out to assume that an inefficient layout should continue to exist. Often an existing space has items that are not really needed. Its contents may have become a constraint for the activities that take place in it. Just moving the old space to a new location, even in concept, could result in serious misfits. In other words, a thorough analysis of people, equipment, activities, and schedule (PEAS) must be completed in Step 1 to help minimize errors in Step 2 and, particularly, in using the comparative method for sizing spaces.

Define Shared Space Requirements (Task 7)

The purpose of this task is to establish the demand for shared- or common-use spaces, convert that demand into a list of shared spaces, and define user requirements for them. *Shared spaces* are rooms, work areas, or other forms of space that serve several or all organizational units.

Shared spaces can be indoors or outdoors. They are identified by the fact that they are not managed or controlled by anyone in particular. They can also be those that belong to one organization but have a poor utilization rate unless others must also use it to increase utilization to an acceptable level.

Examples of commonly shared spaces are found in Table 6-2.

Some shared spaces are primarily used by visitors or the public. User requirements for these spaces must also be developed. It is up to the coordinator to establish what spaces are required, size them, and formulate user requirements for them.

The best way to handle shared spaces in URM is to treat all shared spaces as if they belonged to a nonexisting organizational unit, called "Shared Spaces." As such, a mission and function analysis (Step 1), including PEAS, should be completed. Tasks 1 through 5 in Step 2 should also be completed for the shared spaces.

Using Form G, each representative must estimate demands for each type of shared space needed. The frequency and duration of use for each kind of activity are listed. *Frequency* is the typical number of hours per day, week, or month in which the activity takes place. *Duration* is how long each activity lasts when it does occur. The last data item for demand is the number of people who are involved in the activity. The number of people should normally reflect the usual attendance or participation. However, it may also be important to submit the maximum attendance or participation. While few buildings or rooms can be built for a maximum attendance, the project may require planning for the worst case. The coordinator will have to inform representatives whether typical or maximum attend-

Table 6-2. Examples of shared spaces.

Meetings, Conferences, Training	*Rest, Waiting, Visitors*
Conference room	Lounge
Classroom	Break room
Seminar room	Reception area
Auditorium	Waiting room
Training room	Lobby
Food Service	*Exterior*
Dining room	Parking area
Snack bar	Picnic/Lunch area
Vending room	Waiting area
Lunch room	Passenger loading zone
Cafeteria	Loading docks
Coffee bar	
Hygiene	*Miscellaneous*
Washroom	Copy machine area
Locker room	Reproduction work area
Shower room	Drafting area
Drinking fountain	Maintenance area
	Trophy/Display area

Figure 6-14. Example of a completed Form G.

SHARED SPACES

URM form G

organizational unit FIRE DEPARTMENT

space type	how often used	normal use time	usual and maximum number of people
CONFERENCE ROOM	6 HRS PER WEEK	1 HR (NOTE 1)	5/15
CLASS ROOM	2 HRS PER WEEK	1 1/2 HRS	20/25

comments 1. CONFERENCES: ABOUT HALF LAST 20-30 MIN, THE REST ARE USUALLY 1-2 HRS LONG

ance, or both, should be reported on Form G. Values should be clearly marked as typical or maximum. An example of a completed Form G is illustrated in Figure 6-14.

Figure 6-14 shows that the fire department regularly conducted training. Part of the training was hands on and conducted in fire station bays, shop, and other spaces. Some training required classroom or conference room space to show films, hold strategy sessions, and hear speakers. Classroom and conference-room space could be shared with other organizations in the village.

The fire department estimated that a conference room would be needed about 6 hours per week. About half of the sessions lasted an hour to 2 hours; others were shorter. The usual number of attendees was 5, but sometimes there could be as many as 15. They needed a classroom for about 2 hours per week, with minimum blocks of 1½ hours each time. Typical attendance was 20, while maximum class size was 25 people.

After demand data are submitted by representatives to the coordinator, the coordinator must evaluate the data to determine the number of spaces of each type that are required. Room size is based on the capacity required, a utilization factor, distribution of space types, and other considerations. An example of an analysis of conference room requirements is presented in Figure 6-15.

The coordinator began by making a list of all demands for conference rooms. For conference rooms that have a random demand, a 50 percent utilization factor was known to be reasonable. Looking through the list, Unit H needed a conference room about 30 hours per week. If that unit was

Figure 6-15. Example of the analysis of conference room requirements.

SUMMARY OF DEMANDS

Type of Space	Organization	Use Rate	Typical Attendance
Conference Room	Unit A	5 hr/week	5
	Unit B	10 hr/week	4-6
	Unit F	6 hr/week	3
	Unit H	30 hr/week	8

ANALYSIS

Room Requirements and Estimated Utilization

Unit H should have its own conference room. Utilization is 75 percent (30 hr/week divided by 40 hr/week maximum use = 0.75.)

Units A, B, and F should share a conference room. Utilization would be 53 percent (5 + 10 + 6 hr/week divided by 40 hr/week = 0.53).

Both rooms should be sized for an 8-to 12-person table.

given its own room and properly scheduled its use, it would have a 75 percent utilization factor, based on a 40-hour week. The room should be sized for at least eight people, since that is the typical attendance. *Typical* infers *average. Average* means that 50 percent of the time attendance will be larger. For planning purposes the room should probably be sized for 10 or perhaps even 12 people. Had peak attendance data been provided, the decision would be easier to make.

The remaining three organizational units, A, B, and F had a total demand for a conference room of 21 hours per week (5 + 10 + 6). Based on a 40-hour week, this would yield a 53 percent utilization factor, an adequate demand for a second conference room. Since the maximum typical attendance for these three units was 6 people, the room should be sized accordingly. Again, making it a little larger, perhaps for 10 people, the room would handle above-average attendance.

To determine how much space is needed for different kinds of shared spaces, one can refer to planning and design books, architectural-standards books, and similar sources. One of the four methods for sizing spaces discussed in Task 6 should be suitable.

To avoid low utilization, multifunction spaces can be created. For example, a meeting room can also be used for training sessions. An assembly room might be combined with a gymnasium.

Another way to improve utilization and minimize the amount of space required is to subdivide large spaces into small ones using moving or folding partitions. For example, there may be a low demand for a large conference room and a much larger demand for small ones. A folding partition that doesn't transmit too much sound can be opened for those few occasions when a large room is required. When closed, two small rooms are created. Be sure to identify sound control in Task 5 (Form E) if that is an important requirement for the moving wall or partition.

Utilization factors may be difficult to establish. They are based on demand relative to available spaces during reasonable use hours. What a reasonable utilization factor is will differ, depending on whether demand is random or carefully controlled and scheduled. If activities, such as classroom training, can be carefully scheduled, a utilization factor of near 100 percent is valid. If requirements for a space are not carefully managed and each using group sets its own schedule, a utilization factor of 50 percent may be reasonable for minimizing the likelihood of conflicting schedules. If one organization must have a shared space available on demand, then utilization is irrelevant. The space can only be used by some other organization when the first is not using it. The coordinator must have an understanding of these demands and how shared space will be managed in setting a utilization factor and deciding how many shared spaces are needed.

Some spaces, like dining rooms, will have a peak load. As a result, one cannot achieve a high utilization factor, except for a short time. However, the peak can be reduced considerably through management of lunch times.

Another kind of shared space is washrooms. The number of fixtures and the amount of space can be estimated from the population in the building. The representatives participating in URM should submit to the coordinator the number of males and females included in the personnel identified on Form A in Step 1. Then the data listed below can be used to make an estimate of the total washroom requirements. Distribution of washrooms is usually handled later, during design. One exception to consider is peak washroom demand, such as occurs for schools between class periods or at places of assembly during breaks in activities. The data below does not account for peak demand.

The space required for washrooms depends on a) the type of building, b) the number of occupants in the building, c) the number of fixtures required per occupant, and d) the space required for each type of fixture.

National plumbing codes (such as American National Standard Institute A40.8), or state and local plumbing codes will govern how many fixtures are required for a type of building and the expected number of occupants.

The approximate amount of space required for common fixtures is as follows:

Type of Fixture	*Space Needed per Fixture (sq. ft.)*
Water closet	30
Urinal	16
Lavatory	18
Towel Dispenser (1 per 3 lavatories)	10
Shower	40
Bathtub	26

In Figure 6-16 the total building population is determined by adding up personnel listed on each Form A. Here there were 250 total, 150 males and 100 females. From the local plumbing code it is determined that a minimum of 4 water closets, 3 urinals, and 6 lavatories are required for males and 5 water closets and 5 lavatories are required for females. Using the space standards for fixtures cited above, it is determined that a minimum of 296 square feet is needed for males and 260 for females, or 556 square feet for washrooms.

Analyze the Future (Task 8)

At first when a new building is occupied, it provides a good fit for the occupants. Over time, however, the activities, personnel, equipment, and schedule of the occupants will change. Regardless of what reference time is used in URM, some changes beyond that time can be anticipated. If likely changes are noted, together with the resulting effects on requirements, then a designer may be able to develop a solution that lasts longer and does not become obsolete as soon.

In this task, the goal is to have each representative identify what may change in the future and to suggest the effects these changes may have on the user requirements listed in previous tasks. One should look for changes in organizational structure, missions and functions, and in activities, personnel, equipment, and schedules. Usually a representative will not be aware of many changes. Those higher in the organizational structure are more likely to know about possible future changes. Marketing staff and corporate planners may be good sources of information about future changes.

The effects on buildings cannot always be identified by people at high levels of the organization. Many effects can only be recognized by people at the working level. Representatives should review possible changes with the people in their organizational units to obtain their help in pinpointing effects.

After known or likely changes are identified, they are logged on Form H together with the date when the change is expected to be implemented. Form H also requires that the effects of the changes be listed. The effects indicate how user requirements will change as a result of the changes in PEAS. An example of a completed Form H is found in Figure 6-17.

In this example the police department knew that special radio equipment would be standardized for the whole region in about a year. The equipment would be located in the dispatch room and require an additional 10 square feet. A third shift would probably be added to the support staff, based on the growth of the village. An additional 100 square feet of space would be needed for offices.

In some cases the coordinator, together with top management, may have to deal with change after URM is completed. Information about change may be sensitive, involve the security of the organization, have serious effects on the organization's structure or size, or for other reasons should not be considered by representatives.

Figure 6-16. Analysis of washroom space requirements.

Building Population:		
	Males	150
	Females	100
	Total	250

MALE

Fixture Type	Number of Fixtures	Sq. Ft. per Fixture	Total Area
Water Closet	4	30	120
Urinal	3	16	48
Lavatory	6	18	108
Towel Dispenser	2	10	20
		Total	296

FEMALE

Fixture Type	Number of Fixtures	Sq. Ft. per Fixture	Total Area
Water Closet	5	30	150
Lavatory	5	18	90
Towel Dispenser	2	10	20
		Total	260
		Grand Total	556

Figure 6-17. Example of a completed Form H.

FORECAST

URM form **H**

organizational unit POLICE DEPARTMENT

nature of change	impacts on: type of space amount of space requirements	when expected
NEW RADIO COMMUNICATION SYSTEM FOR ENTIRE COUNTY	10 SQ. FT. EXTRA SPACE IN DISPATCH ROOM EXTERIOR ANTENNA (30 FT. HIGH)	IN 1 YEAR
ADD 3RD SHIFT TO STAFF	100 SQ. FT. OFFICE	IN 3-4 YRS.

Establish General Relationships (Task 9)

In Task 3, the special relationships within organizational units were established. The goal of this step is to define relationships among organizational units and for the total collection of occupants. This is accomplished through group meetings of representatives. Initially, brief meetings are held at one level in the organization above the lowest organizational unit. Several meetings will occur at this level, since each meeting involves related units. For example, assume there are three branches with four sections in each and sections are the lowest units. Each branch will hold a separate meeting with its sections.

Then, meetings are held at each higher level until one final meeting considers relationships among all organizational units. For example, assume there is a department with three divisions and several branches and sections in each division. After each branch has met with its sections, each division will meet with its branches and sections and, finally, the department will meet with divisions, branches, and sections. The low-level units (sections) may be represented by branches if the group becomes too large.

Some judgment is required on the part of the coordinator in determining how many meetings are required and who should be involved at each. A close look at the organizational chart will help in planning these meetings. Figure 6-18 gives another example.

In Figure 6-18 an organization contains three line divisions, each with a different organizational structure. It also contains three staff-support units, each with differing structures. At the first level, a meeting to work out general relationships is required by Division A and Division C. Similarly, Support Unit A and Support Unit C must hold a meeting. Then one final meeting involving top management and all other units is needed to establish relationships among all units of the organization.

Some methods of facility planning require that a space-relationship matrix be compiled for all spaces in a facility. Such a chart is not only difficult to prepare, it is difficult to use, primarily because of its size. The majority of relationships in a large chart are unimportant. Each space will typically have a very limited number of space relationships that are significant relative to spaces in other units. Only a few relationships can be resolved completely for any one space in developing a design.

In URM relationships among blocks of space are considered instead of relationships among all individual spaces. Each *block* is composed of the spaces required by an organizational unit. Relationships among organizational units (and thus blocks) are charted. Only a few key spaces will be referred to individually for relationships among organizational units.

Each meeting is conducted by one of the representatives involved. The meeting leader should be selected by agreement of the attending representatives. Usually it will be headed by the representative of the highest ranking organizational unit. If selection of a meeting leader cannot be agreed upon by the participants, the meeting should be led by the coordinator of URM or someone designated by the coordinator. The meeting for defining relationships among all organizations should be headed by the coordinator. When the meeting is concluded and an organizational-relationship diagram has been completed, a copy should be turned in to the coordinator. Information may be needed in later meetings at higher levels.

A blackboard should be available at these meetings. It is the best device for working out general relationships. Changes in the diagram being formulated can be made easily. It can be copied on paper after the meeting is concluded.

The first step in defining relationships is to determine which organizational unit(s) or space(s) is central to all others being considered. This organizational unit is represented by a circle and placed near the center of the blackboard. The circle is labeled. Then the group must decide which organizational units should be close to the central unit. A circle is added for each unit and labeled. If adjacency is important between two units, circles are drawn so they are touching. If adjacency is not important, some distance should separate circles. The farther apart circles are placed, the less important is the spacial relationship between them. One can also put connecting lines between circles and rate their

Figure 6-18. Meetings are required to establish organizational relationships.

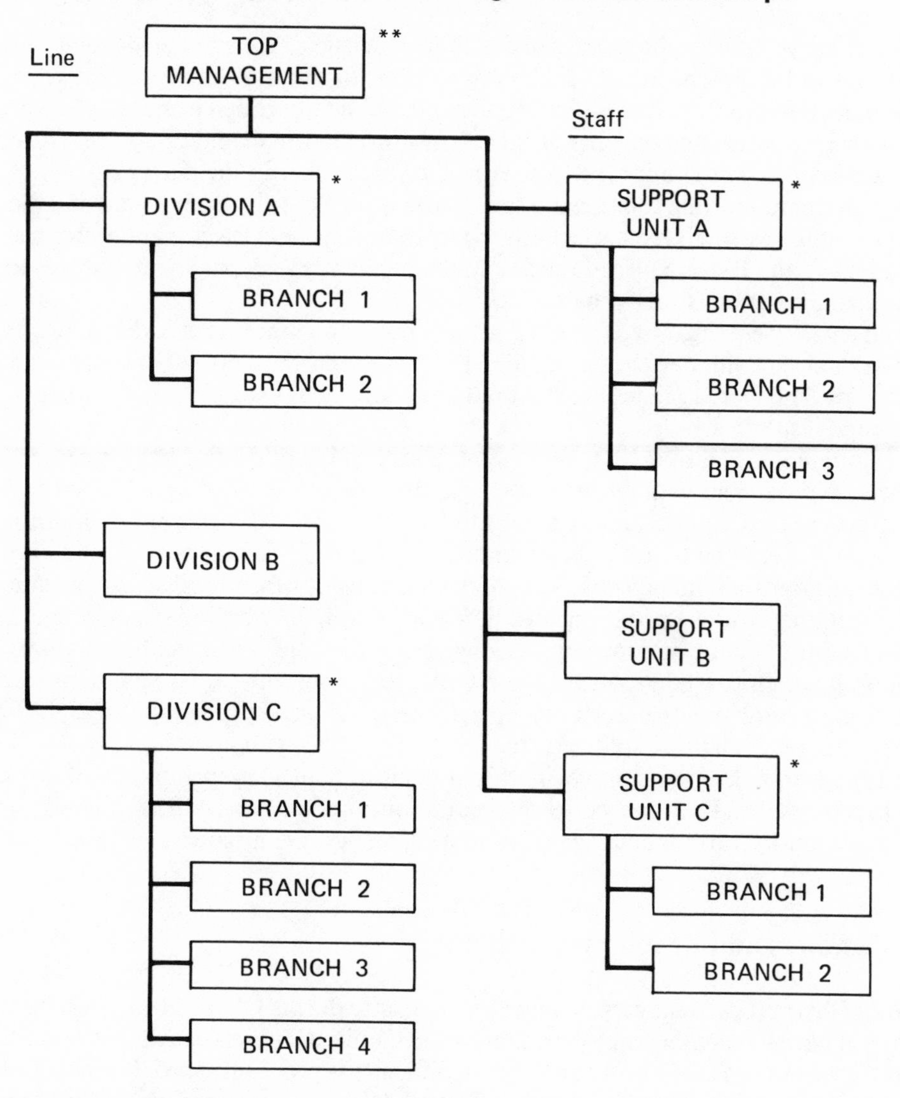

* First level meetings
** Second level meeting (first level must be completed previously)

proximity similar to the procedure for Form C. Figure 6-19 illustrates how general relationships were defined for one project.

For the village of Tolono, the group decided that the village office and the mayor's office were the most important and should be located in the center of the diagram. A circle was drawn and labeled. Each of the other organizational units (fire department, public works department, police department, and ESDA) were placed around the central circle. The fire department and ESDA were related to each other during certain activities and were placed next to each other in the diagram.

The village board meeting room was also important for the village office staff and the mayor. It was placed near the center circle as well. Other spaces and facilities (sewer plant, water plant, garage, and fire stations A and B) were placed relative to the public works department and fire department. The garage and fire station B was a shared facility.

After all circles were organized in the diagram, connecting lines were added to clarify relationships. For example the fire department and ESDA both used the village board meeting room for training sessions. The shared garage/fire station B was connected to the public works department and fire department, respectively.

As is illustrated in Figure 6-19, it may be necessary to include a few key spaces in the overall diagram, representing them with circles as well. Certain spaces (usually shared by several organizational units) may be important in understanding the general relationships among organizational units.

Remember, general-relationship diagrams are not intended to define spacial relationships for all spaces. These diagrams will mainly chart functional relationships among organizational units. This is particularly true the further up the organization one moves. In working out a design solution, a designer may wish to define relationships among spaces in more detail. As noted above, the large space-relationship matrices (an expanded Form C) are then appropriate. The diagrams might also be constructed from a combination of relationship matrices developed by individual organizational units and the organizational relationship diagrams.

The meetings among representatives in which organizational relationships are defined can also be used for other purposes. They can be used to communicate how well URM is proceeding and to resolve conflicts among certain organizational units. Conflicts are discussed in Task 10 below.

Resolve Conflicts and Problems (Task 10)

The purpose of this task is to take care of difficulties that arise during URM. There are at least two types of difficulties. The first is conflicts among organizational units. Conflicts usually result from potential changes to the way things are. The second type of difficulty is procedural problems in implementing URM and completing URM steps and tasks efficiently. Some of the conflicts and problems can be foreseen, others cannot.

Dealing with Conflicts

Planning for the future and threats of change are very frightening for people. Most are satisfied with the status quo, because they know where they stand, be it in the organization, with their peers, or in terms of rank and pay. Because planning for buildings requires an evaluation of the way things are and recommendations about the way they should be in the future, participants may note or create conflicts. These conflicts involve such things as individual roles, roles and status of organizational units, changes in staffing, authorization for new equipment or activities, and control of space.

Many of the conflicts that arise out of URM or any other planning method can only be solved by someone at a level higher than at the level where the conflict occurs. The coordinator cannot solve all

Figure 6-19. Example of an organizational relationships diagram.

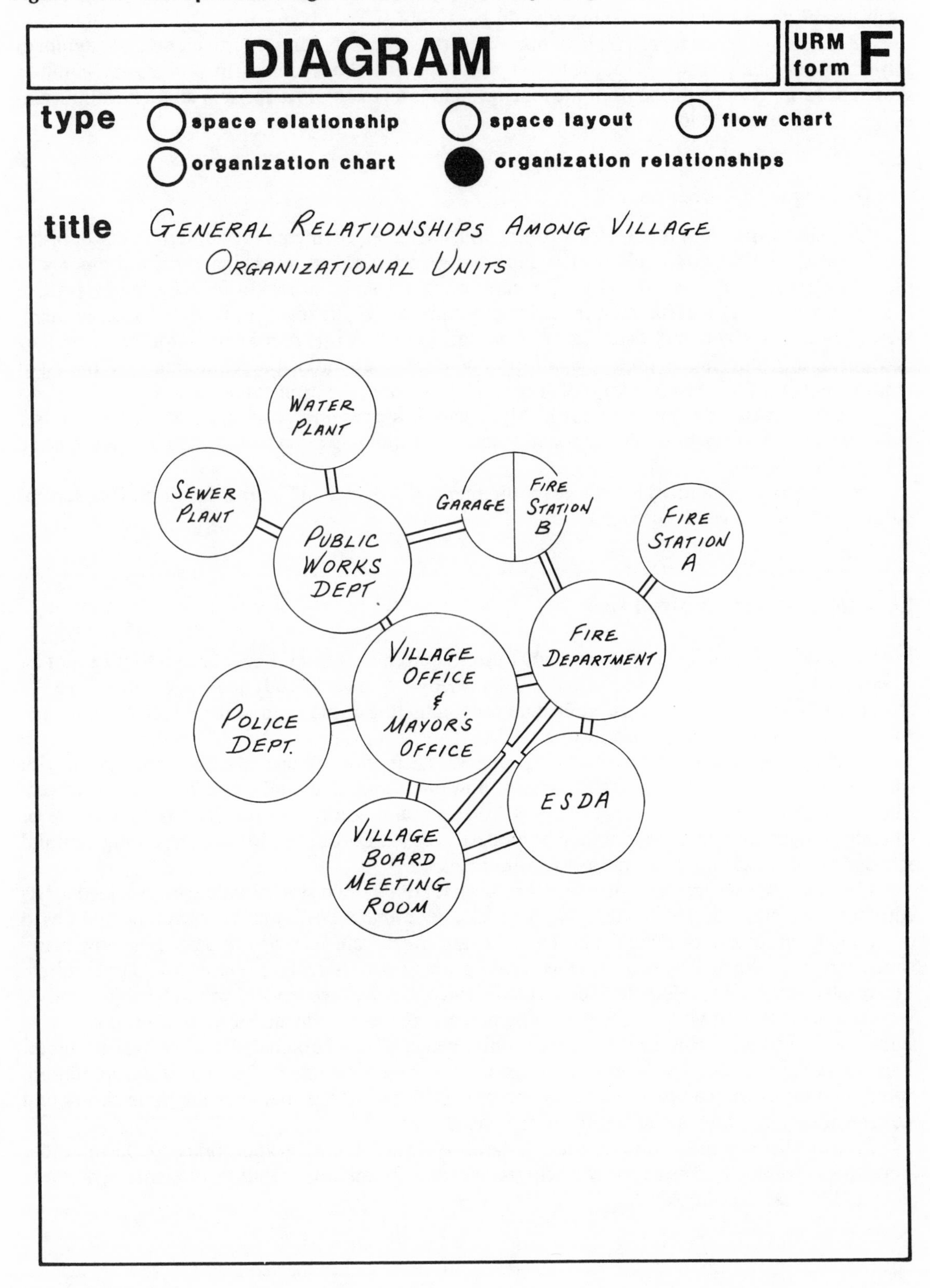

the conflicts, and neither can the representatives. Very often, top management must be brought in to resolve conflicts and decide how things should be.

The coordinator and representatives must be alert for conflicts among participants and attempt to solve them as quickly as possible at the lowest possible level of the organization. If necessary, conflicts should be reviewed with the appropriate management level to obtain a decision about how information in URM should be logged.

Dealing with Problems

One kind of problem that arises is having to list user requirements for numerous spaces of the same type. Obviously, one would want to reduce the work required in logging user requirements by minimizing redundancy in data. Form E is most often affected. For repetitive spaces and repetitive user requirements, a list of standard requirements is prepared. The spaces or types of spaces to which they apply are noted with each list. When representatives itemize user requirements for the spaces they need for their organizational units, they simply reference a standard list of requirements or the list of requirements applicable to that kind of space and note only exceptions or additions.

The coordinator must be alert for highly repetitive spaces and develop a standard list of user requirements. There might be one list for all spaces in the building or several lists, each applicable to a different kind of space.

An example of a standard list of user requirements is provided in Figure 6-20. Form I is a general purpose form which can be used here.

Conduct Special Studies (Task 11)

In some cases the need for a space and what user requirements should be attached to it cannot be determined from past experience or current knowledge. A special study may be needed. Task 11 recognizes that possibility. General guidelines for conducting a study cannot be made. An approach will have to be selected by those applying URM.

In many cases, studies are used to determine what activities will take place in a building or with what frequency. They may be used to determine how many and what kinds of people will be involved. They may also be necessary to determine whether a major equipment purchase is appropriate or whether an aging equipment item should be modernized. Studies can also be made regarding time and scheduling that could influence project requirements.

For example, one organization was not sure what kind of food service should be provided for its employees in a building. The building was to be located several miles from any restaurant, and was to be in operation for two or three shifts. The 300 employees were surveyed to determine how many would carry their lunch to work or whether vending, hot-grill, or full dining service would be required. The results were used to select the kind of food service that was needed and would be cost-effective.

Another organization conducted training programs at various locations around the country. A study was done to determine if a centralized training center should be included with a new home office. The number of classes and students for the last few years were tabulated. The cost of transportation, hotels, and meals was compiled. A projection of cost and time was made for the same classes and students if they had been located at the home office.

In cases where special studies are needed, the use of an expert may be appropriate for the topic that is creating the difficulty. The expert can help resolve the information, operation, or planning problem so that URM can be completed.

Figure 6-20. Example of a standard list of requirements for office spaces.

URM form I

REQUIREMENTS ASSUMED APPLICABLE TO ALL OFFICE SPACES
for
MUNICIPAL BUILDING

ACCESS
- Closing door for private offices
- Maximize visual privacy

ENVIRONMENTAL
- Thermal comfort conditions, including humidity control in winter
- Sound transmission control for private offices
- Lighting for reading

UTILITIES
- 110-volt, single-phase receptacles for variety of office machines and equipment

APPEARANCE
- Pleasant decor—color and texture coordinated with one of three standard schemes

FINISHES
- Low maintenance
- Fire resistant materials

Figure 6-21. Sample memorandum which provides instructions to representatives about verifying Step 2 data.

MEMORANDUM

To: All Representatives Participating in URM

From: David Paul, Coordinator for Project X

Subject: Verifying Step 2 Data

1. It is strongly suggested that each representative have coworkers in your organizational unit and the head of your unit review the requirements and supporting data prepared in Step 2. Soon you will be preparing your data for submittal. You will not be able to make changes after you turn them in. Check your information now.

2. One approach is to make a copy of your data sheets for each member of the group, send it to him or her with instructions for reviewing it and making comments. You should have all comments back and corrections completed by July 10. Another way is to walk everyone in your unit through the data sheets in a meeting. You can explain each form. The resulting discussion may surface some key oversights or errors.

3. Have reviewers look for a) missing requirements and data, b) unnecessary requirements and data, c) additional future changes, and d) any errors.

Verify Data (Task 12)

Similar to the last task in Step 1, there is a need in Step 2 to verify data logged in URM. People other than the representative within an organizational unit may identify details that were overlooked or logged in error. Comparing data across organizational units as URM progresses upward through the organization will uncover inconsistencies in data and conflicts among organizations. These difficulties can be resolved in Task 10.

This task simply recognizes that one person cannot know all requirements. Representatives are merely spokespeople and must draw on the knowledge of others whom they represent to ensure that logged data are correct. Data should be reviewed within a unit and by at least one level above a unit.

A sample memorandum in Figure 6-21 explains what a representative wants coworkers to look for in reviewing data for their organizational unit.

Summary

Step 2 is the heart of URM: the requirements are formulated here. The step builds on the analysis of mission, functions, and PEAS in Step 1. It involves twelve tasks. The first nine are completed in a sequence. The latter three are fit into the sequence in different ways, depending on project need.

In this step spaces are named, what will be in them is established, how they should be spacially related is determined, and what requirements and features they should have is defined. The size of each space is estimated using one of various methods. Demands for shared space are used to determine how many and what kind of shared spaces are needed and what features they must have. Changes foreseen beyond the reference time for the project are estimated together with effects on requirements. Overall relationships among organizational units are defined in group meetings and remaining issues and problems are resolved. If necessary, special studies are completed to establish the need for certain spaces and requirements.

Chapter 7

Communicating User Requirements

Once participants have identified and logged user requirements, they will undoubtedly feel like they have completed their task. The goal was to define what is needed. While that goal has been met at this stage, URM has not been completed. User requirements must be communicated and applied. Otherwise URM is a waste of time. In order to communicate user requirements, they have to be organized and documented in a way that will make it easy for someone who doesn't know much about them to find the requirements and apply them in a solution.

The goal for this step is to organize the user requirements and supporting data in a manner that will foster their communication and use. To accomplish this goal, Step 3 is divided into four tasks: The first task is preparing the organizational unit data; the second is preparing the summary data. The third task is formulating the background data, and the fourth is completing the URM document.

In Task 1, each representative prepares a final document using data of the organizational unit. The coordinator does the same for shared-space data. In Task 2, a summary of key data is prepared by the coordinator. It forms an index to the data of each organizational unit and gives total requirements for space. The objective for the project, list of occupants, and a brief description of organizational units and overall operations are compiled by the coordinator in Task 3 in a section of the final document called "Background Data." It will introduce readers to the document and explain why it was produced, and who it describes. In Task 4, all final material is prepared for a finished document and reproduced.

There is no single way to organize user requirements and supporting data into a document that will allow a designer or someone else to look up information easily in all the ways necessary to apply it. The format presented in this step is just one way of documenting URM data.

Computer files and data-base management systems can also be effective means for compiling, storing, and querying URM data. Almost any combination of data can be looked up quickly. Final documentation can be created if printed output is needed. See Appendix B for more information about using the computer for managing URM data. If data is to be compiled in a data-base management system, forms completed in Steps 1 and 2 of URM can be used to load the computer files. Much of Step 3 can then be performed by the computer after appropriate reports are set up.

The coordinator must instruct representatives when and how to prepare and submit data. Typing may be distributed among organizational units or handled centrally. A variety of specifications may be needed, depending on how data will be compiled. The discussion in this chapter assumes that a document is being produced, rather than loading data into computer files. A format known to be effective is described below, although some adjustments may be desired for particular projects.

Figure 7-1. Sample memorandum instructing representatives about preparation, organization, and submittal of URM forms.

MEMORANDUM

To: All Representatives Participating in URM

From: David Paul, Coordinator for Project X.

Subject: Instructions for Preparation, Organization, and Submittal of Final Forms

1. This activity has moved along very well to date. You have all been very cooperative and punctual. So far in URM we have completed the development of user requirements for the proposed new facility. To ensure that they are used in further planning and design, it is very important that we make them readable, consolidate them, and organize the final document well.

2. As we discussed in our meeting about Step 3 of URM, please transfer all your requirements and supporting data to copies of Forms A, F, H, J and K. Your final forms must be typed.

3. Place your data sheets in the following order for submittal:
   Form A (Mission/Functions)
   Form F (Space Relationship Diagram)
   Form H (Forecast)
   Form J (Space & Requirements)
   Form K (Contents of Space)
Add a Form J and Form K for each space that you have identified for your organizational unit. If you have 7 spaces, there should be 7 sets of Forms J and K attached in their order of the space ID numbers you assigned to them.

4. I must receive your original forms not later than July 25. Those received late will be counted toward delinquent performance in URM. Be sure to make a copy for yourself.

5. If you were responsible for preparing an organization relationship diagram (Form F), I should have already received that diagram from you. If I do not have it, it will be noted at the bottom of this memorandum.

Prepare Organizational Unit Data (Task 1)

The information is compiled into final form by each representative for submittal to the coordinator. Data is presented mainly on four forms: A, F, J, and K. A Form H will also be needed if "future" data has been projected for an organizational unit.

The coordinator must explain to representatives how final forms are to be prepared, organized and submitted. Figure 7-1 is a sample memorandum for instructing representatives.

The Organizational Unit

Each organizational unit is characterized in a final document by its mission and functions, its spaces, and space relationships within the unit. Two forms are used: Forms A and F. Together they give a reader a quick understanding of an organizational unit and the spaces it needs.

Mission and function data on Form A has already been completed. It need only be redone to make it neat. Figure 7-2 gives an example of a completed Form A. Instructions for completing Form A were given in Chapter 5.

The spaces required by an organizational unit should be presented on Form F. The form is then marked a "space relationship" diagram. Some form of bubble diagram is an appropriate way to present relationships. Figure 7-3 illustrates one possible scheme for a bubble diagram. A circle is used to represent each space. A line or bar connects spaces, and the value of relationships previously entered in the matrix on Form C is located on the lines or bars. Detailed instructions for preparing bubble diagrams are found in Appendix C.

The reason matrix data is converted into bubble diagrams is to foster communication. A graphic representation may be more effective than the matrix of data. The greater the number of spaces included in a bubble diagram, the more difficult it is to include all the relationships from a matrix, like that prepared on Form C. When there are many spaces and complex relationships, only important relationships are included in a bubble diagram. If the bubble diagram gets too complex and visually busy, the list of spaces and matrix from Form C can be included with the bubble diagram on Form F (see Figure 7-4) or a copy of Form C can be included (see Figures 6-2 through 6-4).

The Forecast

Forecast changes for an organizational unit and effects on requirements were listed on Form H. The changes were projected to occur some time after the reference time that was selected for the URM application. No changes are needed in previous data unless necessary to make Form H neat. A completed Form H that suggests changes and impacts on requirements is found in Figure 7-5. Instructions for preparing Form H were given in Chapter 6 (see Figure 6-17).

The Requirements for Spaces

One Form J is completed for each space. Requirements for a space are listed on it in final format. Data is transferred from Form E to Form J. If a layout sketch was prepared on a Form F, the sketch should also be transferred to Form J. Figure 7-6 shows a completed Form E and Form F for the same space. Figure 7-7 shows the same data combined on a Form J in final format.

One Form K is also completed for each space. What will be in a space and how it will be used are shown in final format on Form K. Data from Form B is transferred to Form K. One should also refer to the "PEAS" column on Form C to make sure that functions and activities assigned to the space are correctly entered on Form K. Use two Form Ks if there is not enough room on one.

(text continued on p. 118)

Figure 7-2. Example of Form A in final format.

MISSION/FUNCTIONS

URM form **A**

organization

FIRE DEPARTMENT

representative

DAVID GREEN
Ext. 459

personnel

male	10
female	2
total	12

mission

To provide life and property saving service to the community.

functions

A. Fire calls
B. Rescue calls
C. Administration
D. Education and promotion
E. Firemen training
F. Equipment and vehicle maintenance
G. Storage

subordinate organizational units

Village Board
- FIRE DEPARTMENT
  - Fire Brigades
  - Rescue Squad
  - Maintenance Unit

Figure 7-3. One method for formatting bubble diagram.

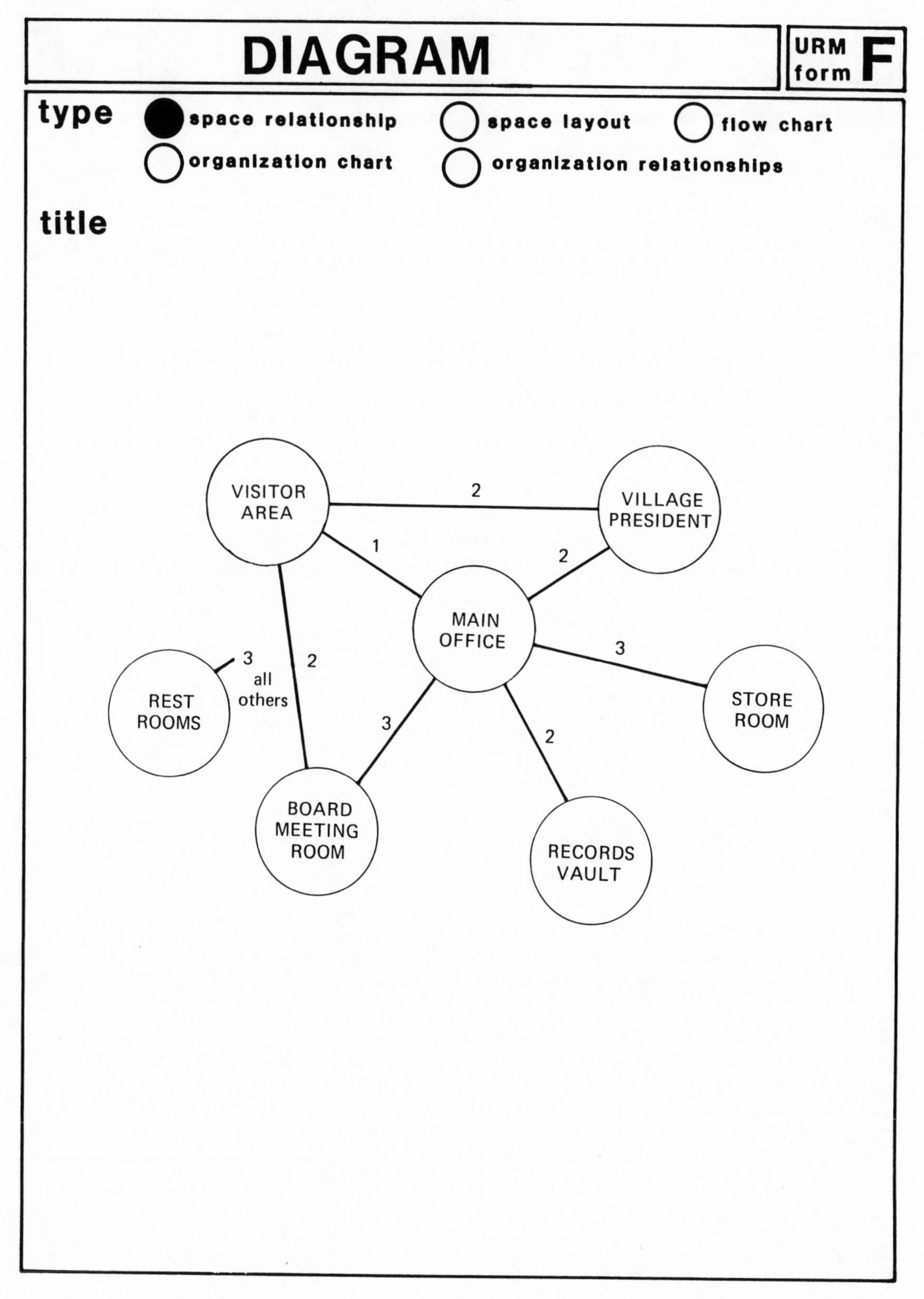
DIAGRAM

URM form F

type

space relationship

space layout

flow chart

organization chart

organization relationships

title

VISITOR AREA

VILLAGE PRESIDENT

MAIN OFFICE

REST ROOMS

STORE ROOM

BOARD MEETING ROOM

RECORDS VAULT

3 all others

Figure 7-4. Relationship matrix added to bubble diagram to preserve all relationship data.

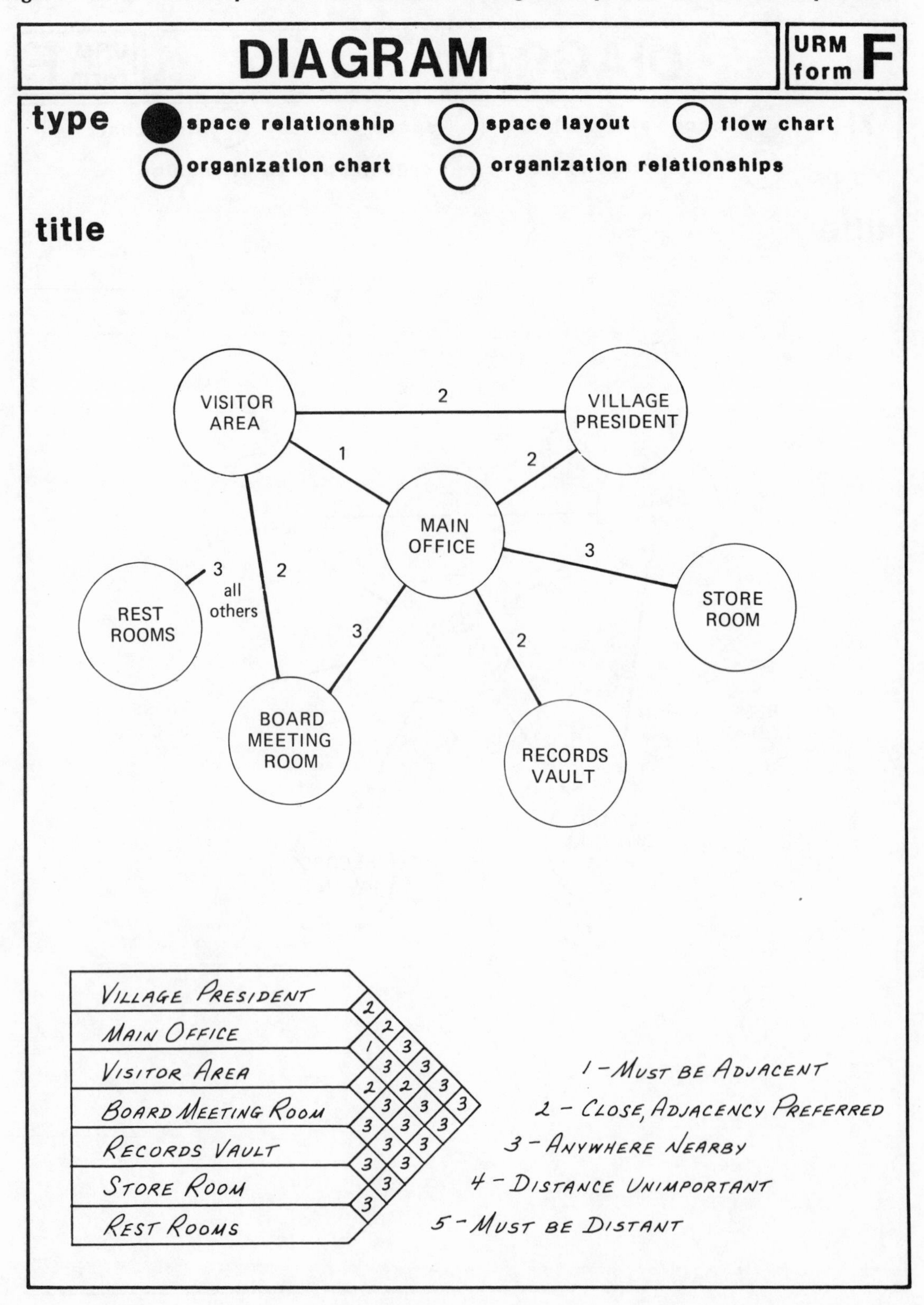

Figure 7-5. Example of Form H in final format.

FORECAST

URM form **H**

organizational unit POLICE DEPARTMENT

nature of change	impacts on: type of space, amount of space requirements	when expected
New Radio Communication System for entire county	10 square feet extra space in Dispatch Room Exterior antenna (30 feet high)	1 year
Add third shift to staff	100 square feet in office area	3–4 years

Figure 7-6. Form E and Form F data that will be transferred to Form J.

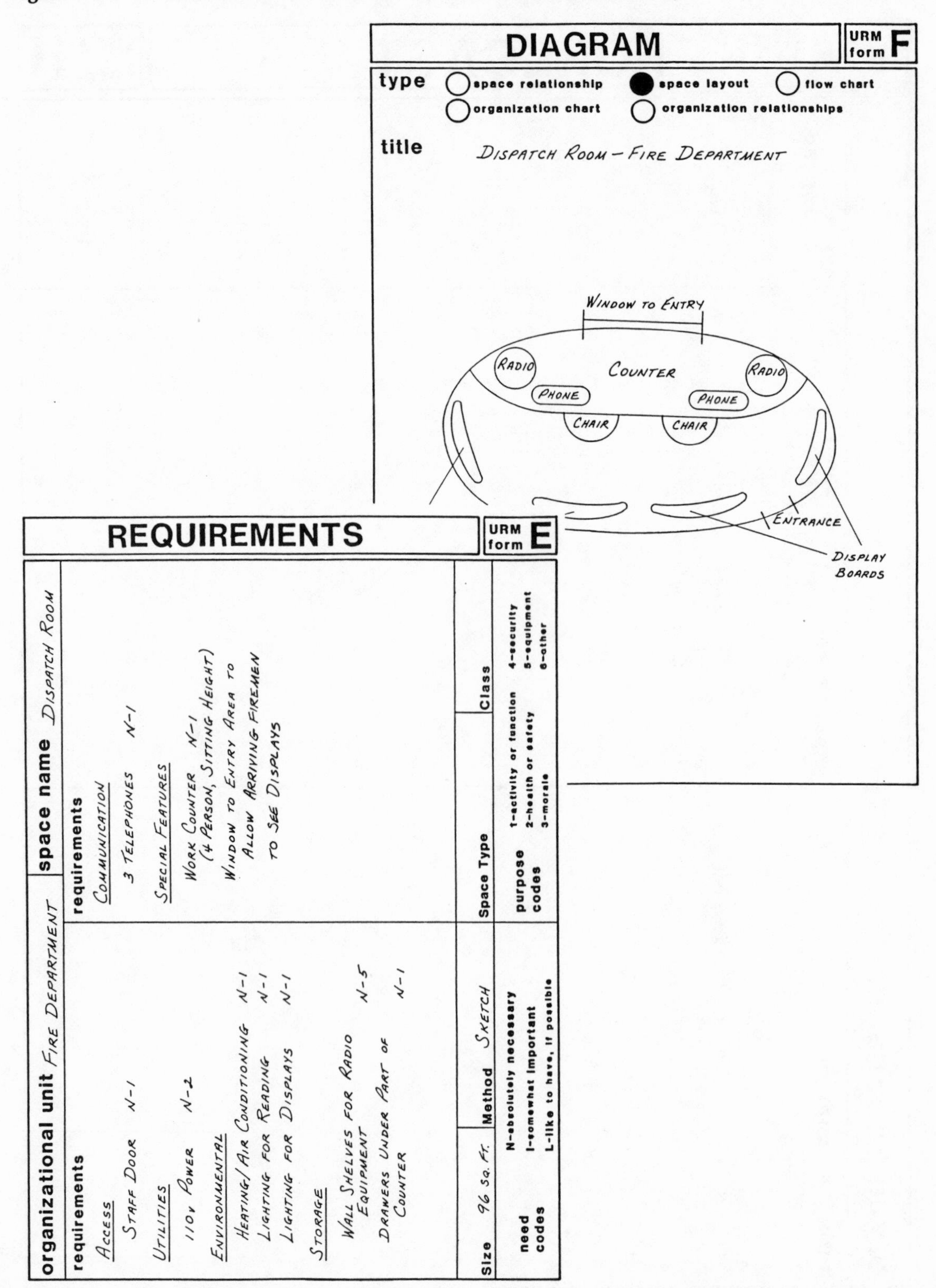

DIAGRAM

URM form F

type: space relationship; ● space layout; flow chart; organization chart; organization relationships

title: DISPATCH ROOM – FIRE DEPARTMENT

REQUIREMENTS

URM form E

organizational unit: FIRE DEPARTMENT

space name: DISPATCH ROOM

requirements

ACCESS
STAFF DOOR N-1

UTILITIES
110V POWER N-2

ENVIRONMENTAL
HEATING/AIR CONDITIONING N-1
LIGHTING FOR READING N-1
LIGHTING FOR DISPLAYS N-1

STORAGE
WALL SHELVES FOR RADIO EQUIPMENT N-5
DRAWERS UNDER PART OF COUNTER N-1

requirements

COMMUNICATION
3 TELEPHONES N-1

SPECIAL FEATURES
WORK COUNTER N-1 (4 PERSON, SITTING HEIGHT)
WINDOW TO ENTRY AREA TO ALLOW ARRIVING FIREMEN TO SEE DISPLAYS

Size: 96 SQ. FT. | Method: SKETCH | Space Type | Class

need codes: N-absolutely necessary; I-somewhat important; L-like to have, if possible

purpose codes: 1-activity or function; 2-health or safety; 3-morale; 4-security; 5-equipment; 6-other

Figure 7-7. Final format on Form J for data from Forms E and F.

SPACE & REQUIREMENTS

URM form **J**

organizational unit	FIRE DEPARTMENT	space name	DISPATCH ROOM

basis for size estimate (method, standards)	size (sq ft)	space number
SKETCH	96 square feet	F-6

requirements

ACCESS
Staff door N-1

UTILITIES
110 v power N-2

ENVIRONMENTAL
Heating/air conditioning N-1
Lighting for reading N-1
Lighting for displays N-1

STORAGE
Wall shelves for radio equipment N-5
Drawers under part of counter N-1

COMMUNICATION
3 telephones N-1

SPECIAL FEATURES
Work counter N-1
(4 person, sitting height)
Window to entry area to allow arriving firemen to see displays N-1

layout sketch

Figure 7-8. Form B and Form C data that will be transferred to Form K.

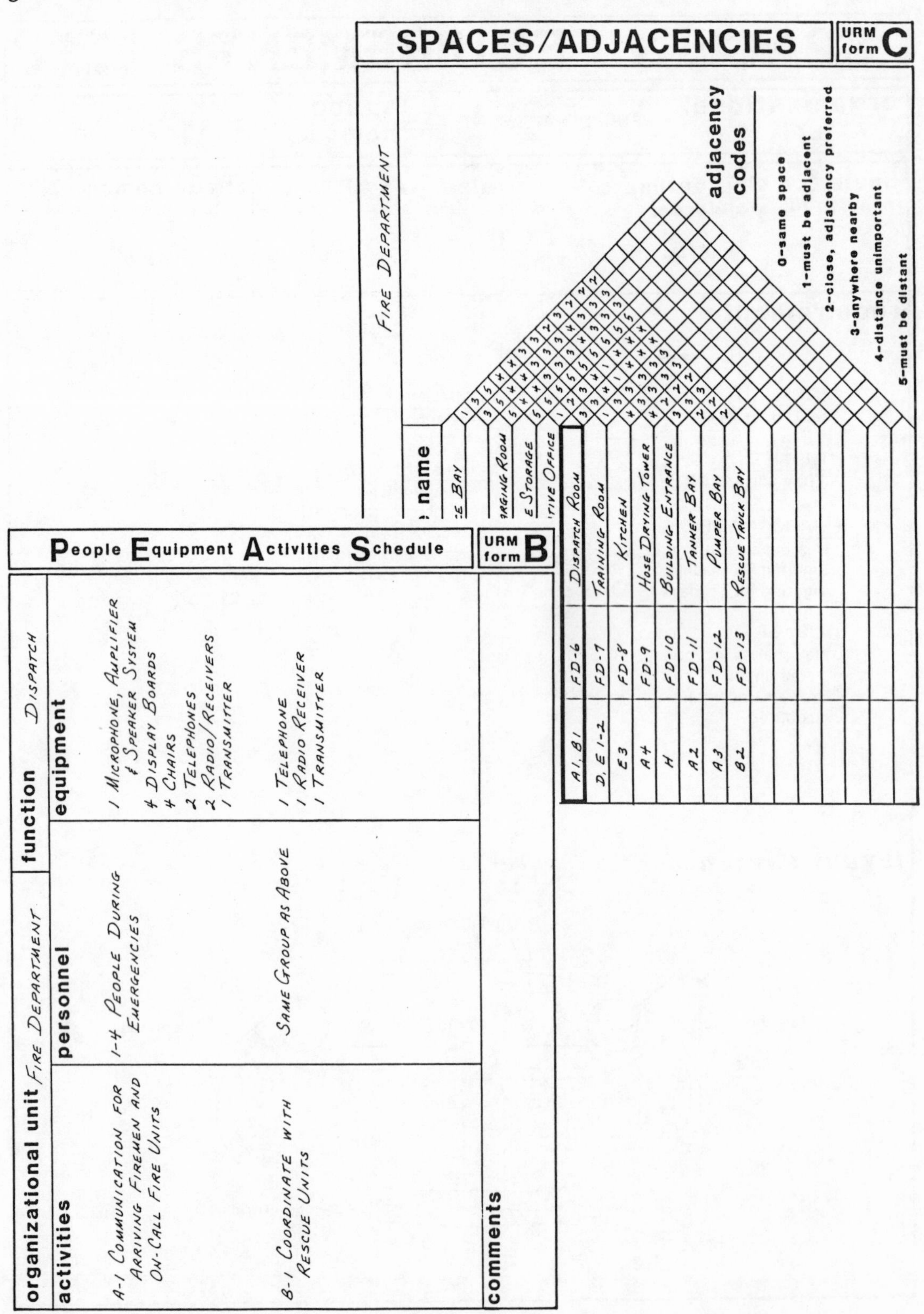

SPACES/ADJACENCIES — URM form C

FIRE DEPARTMENT

name		
...E BAY		
...RGING ROOM		
...E STORAGE		
...TIVE OFFICE		
DISPATCH ROOM	FD-6	A1, B1
TRAINING ROOM	FD-7	D, E1-2
KITCHEN	FD-8	E3
HOSE DRYING TOWER	FD-9	A4
BUILDING ENTRANCE	FD-10	H
TANKER BAY	FD-11	A2
PUMPER BAY	FD-12	A3
RESCUE TRUCK BAY	FD-13	B2

adjacency codes
0–same space
1–must be adjacent
2–close, adjacency preferred
3–anywhere nearby
4–distance unimportant
5–must be distant

People Equipment Activities Schedule — URM form B

organizational unit FIRE DEPARTMENT — function DISPATCH

activities	personnel	equipment
A-1 COMMUNICATION FOR ARRIVING FIREMEN AND ON-CALL FIRE UNITS	1-4 PEOPLE DURING EMERGENCIES	1 MICROPHONE, AMPLIFIER & SPEAKER SYSTEM 4 DISPLAY BOARDS 4 CHAIRS 2 TELEPHONES 2 RADIO/RECEIVERS 1 TRANSMITTER
B-1 COORDINATE WITH RESCUE UNITS	SAME GROUP AS ABOVE	1 TELEPHONE 1 RADIO RECEIVER 1 TRANSMITTER

comments

Figure 7-9. Final format on Form K for data from Form B and Form C.

CONTENTS OF SPACE

URM form **K**

organizational unit	FIRE DEPARTMENT	space name	DISPATCH ROOM

activities

people & equipment

ACTIVITY: A1
Communication for arriving firemen and on-call fire units

PEOPLE
1-4 people during emergencies

EQUIPMENT
1 microphone, amplifier & speaker system
4 display boards
4 chairs
2 telephones
2 radio/receivers
1 transmitter

ACTIVITY: B1
Receive calls; coordinate with rescue units

PEOPLE
Same as Activity 1

EQUIPMENT
1 telephone
1 radio/receiver
1 transmitter

comments

A Form B is illustrated in Figure 7-8. Figure 7-9 shows a completed Form K that is made up of data transferred from Forms B and C in Figure 7-8.

Each pair of Forms J and K combine all data about a space. A reader will not have to search through various forms to gain an understanding of the space and its requirements, use, and contents.

Final Organization and Submittal

After a representative has prepared final forms as recommended here or specified by the coordinator, the forms for the organizational unit should be put into a logical order and submitted to the coordinator. The coordinator's staff will combine all forms into a final document and reproduce it in needed quantities.

Figure 7-10 illustrates how final forms should be organized. Forms A and F, which describe the organizational unit, should be on the top of the stack. If Form C is submitted, it should be directly behind Form F. The forecast data, Form H (if changes were anticipated), should follow. Forms J and K for the first space of the organizational unit follow. Then Forms J and K for the second space, third space, fourth, et cetera should follow, until all spaces are accounted for.

When final forms have been properly organized, they should be submitted in accordance with instructions provided by the coordinator. Representatives may want to retain a copy of all forms completed for URM, particularly those submitted, for backup and later reference.

Prepare Summary Data (Task 2)

Summary data are primarily tabulations or pools of detailed data from organizational units. The tables are derived from the forms submitted by each coordinator. A variety of tabulations are possible. If tabulations contain quantified data, subtotals and grand totals may be included in the table. For example, one table may contain a list of all spaces required by organizational units and their sizes. If the list is organized by units, then subtotals could be computed that state the total amount of space each organizational unit needs. A grand total for all organizational units could be included at the end of the table. Subtotals for each level of the organization might also be included.

Summary data are compiled by the coordinator and his/her staff from data submitted by representatives. Unless data is stored on a computer, the summary tables will have to be prepared by hand and checked for accuracy. Summary data typically include:

1. A table listing all required spaces and their sizes
2. A table listing the quantities of space required by each organizational unit and the total space needed for all units combined
3. The overall relationship diagrams that show the relationships among organizational units
4. A table of forecasted changes and effects
5. Other summary tables as desired

A variety of summary tables is possible besides those based on organizational units. Summaries of space quantities required by kind of space, by major elements of an organization or by function may also be desired. Examples of summary data are provided in Figures 7-11 and 7-12.

In Figure 7-12 all spaces are tabulated. Some organizational units are not shown to shorten the length of the table. Mechanical and circulation space was added near the end of the table. Adjusting net usable space to gross space is discussed below.

In Figure 7-13 all forecast changes and effects are tabulated for an organization and its URM project. Not every organizational unit will have changes for the list, only those completing a Form H.

Figure 7-10. Organization of forms representatives submit to the coordinator.

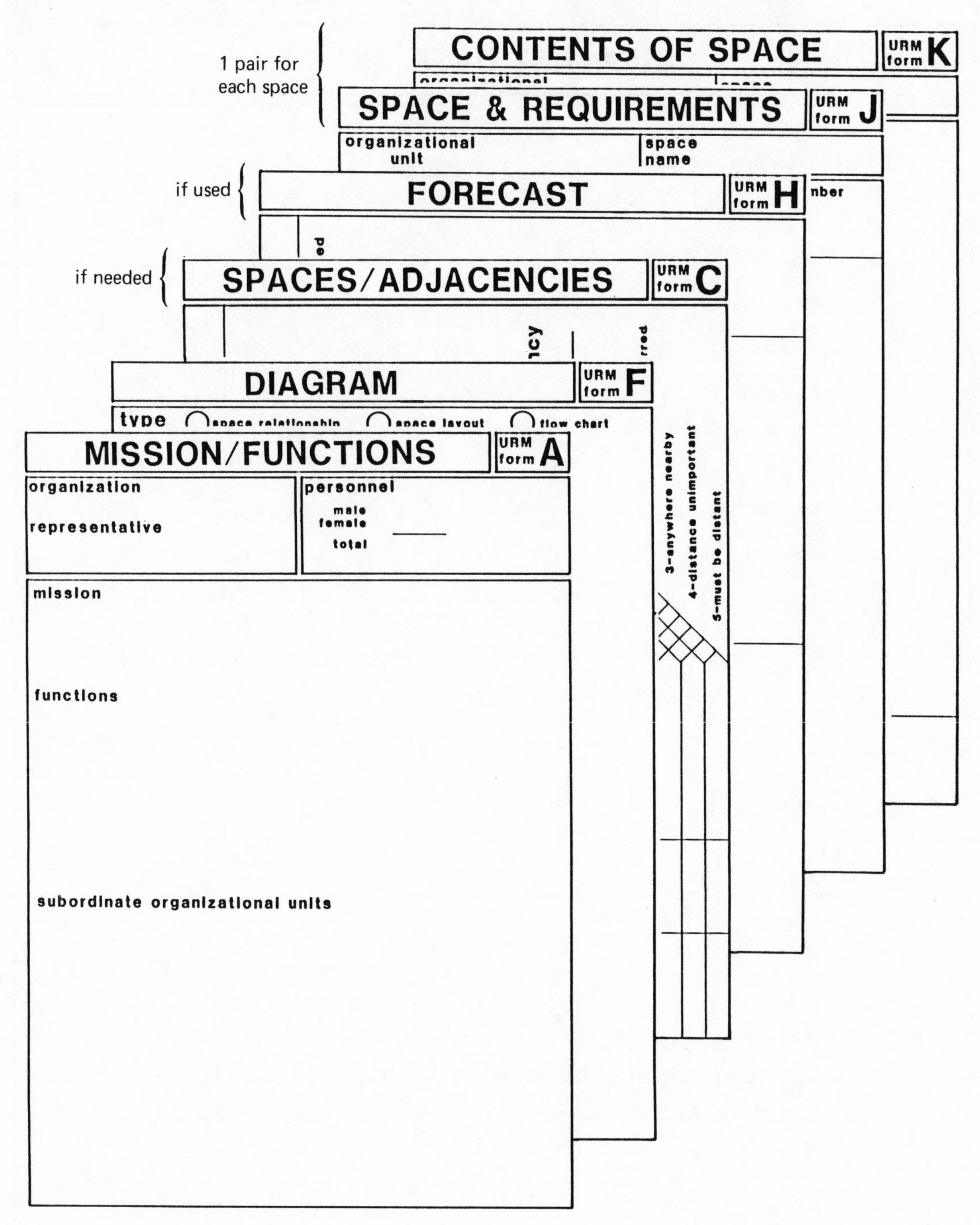

Figure 7-11. List of all required spaces and their sizes.

SUMMARY DATA

URM form I

TABULATION OF TOTAL SPACE NEEDS

MAIN OFFICE		SIZE, SQ. FT.
Reception Area		60
Staff Office		200
Administrator Office		120
Store Room		250
Display Area		80
Meeting Room		400
	Subtotal	1110
DEPARTMENT A		
Adminstrative Office		125
Work Area		30
Storage Room		40
	Subtotal	195
DEPARTMENT C		
(and other Departments through Department F)		
•		
•		
•		
DEPARTMENT G		
Administrative Office		150
Garage		800
Shop		600
Store Room		50
Worker Locker Room		100
Bulk Storage		2500
	Subtotal	4200
TOTAL NET USEABLE SPACE		14,520
MECHANICAL, CIRCULATION (20%)		2,904
TOTAL GROSS SPACE		17, 424

Figure 7-12. List of all forecasted changes.

SUMMARY DATA	URM form I

FUTURE CHANGES SUMMARY

ORGANIZATION	NATURE OF CHANGE	WHEN EXPECTED	IMPACTS ON REQUIREMENTS
Main Office	Additional clerk	2 years	100 sq. ft.
Department A	Add a computer	1 year	300 sq. ft. Air conditioning
Department F	Packaging Machine	1 ½ years	400 sq. ft. for machine 100 sq. ft. storage area for added supplies Special Lighting
Department X	New Dept.	3 years	400 sq. ft. office 200 sq. ft. production area 220 v, 3 phase power 200 gal/min water supply

An overall relationship diagram among organizational units was illustrated in Figure 6-19. Overall relationship diagrams for an entire organization and major elements of it are presented in the summary section.

Net Useable Space versus Gross Space

The amount of space needed by an organizational unit does not include the space occupied by walls. The space required by a user is often termed *net usable space.* When wall thickness and other space not included in "net" space are presented in a table of space requirement, *gross space* is referred to. Some circulation or traffic space is included in both of these definitions of space. However, the main corridors, escalators, or elevators in a building are not included in net space and must be added to the user-space needs before a final estimate can be made of space needed in a building. In addition, some space must be included for heating and air-conditioning equipment and vertical ducts (mechanical space).

A rule of thumb can be used to convert net usable space to gross space that includes wall, circulation, and mechanical space. Although the amount of adjustment required will vary by the type of building, a 20 percent adjustment is reasonable. Later this factor can be modified, if necessary, by a designer to improve accuracy in gross space. At least the difference between net and gross space will not be overlooked entirely. That could cause a severely undersized project. The tabulation of total space needs shown in Figure 7-11 includes an adjustment at the end of the table.

Prepare Background Data (Task 3)

Background data introduces the reader of the final document (usually a designer) to the URM data, the organizational units that the data represents, and why URM was completed. Background data appears at the beginning of the document. This typically includes:

1. The objective for the building or project
2. A list of occupants and numbers of people
3. An overview of the uses, functions, and activities in the building including—
    a. flow, process and organizational diagrams and
    b. brief descriptive paragraphs about organizational units

The background data is prepared by the coordinator and his or her staff. Examples of background data are found in Figure 7-13 through 7-16.

Sample objectives were illustrated in Figure 4-11. An objective was drafted when URM was being organized. One example is repeated in Figure 7-13. The original draft may have been modified somewhat by the time final documentation is prepared.

A draft list of occupants was also prepared while URM was being organized. Figure 4-10 gave examples. One example is repeated in Figure 7-14. Again, the list will often be modified by the time URM is completed.

Flow charts help designers grasp how an organization operates. Flow of paper, materials, supplies, people, vehicles, assemblies, and other things often govern how an organization is structured and what spaces and facility requirements it needs. One flow chart is shown in Figure 7-15.

Another way to help introduce designers or others who must use URM data to the organizational units included is to write a brief paragraph about each unit. Examples for a few organizational units are shown in Figure 7-16.

Figure 7-13. Sample objective statement.

BACKGROUND DATA	URM form I

OBJECTIVE

The objective for this application of URM is to define the administrative space requirements of all elements of the XYZ Corporation located within 20 miles of the current headquarters. The resulting data will be used to determine a) if a new head-quarters facility is needed and b) if so, what elements of the organization will be located in it, what elements will be located in existing buildings that will be renovated and which of the current buildings should be torn down or sold.

Figure 7-14. Sample list of occupants.

BACKGROUND DATA

URM form I

LIST OF OCCUPANTS

ORGANIZATIONAL UNIT	NUMBER OF PERSONNEL	NUMBER OF MALES	NUMBER OF FEMALES
Sales Division	5	2	3
Product X Sales Branch	3	1	2
Aircraft Section	6	3	3
Automotive Section	10	4	6
Trucking Section	12	8	4
Product Y Sales Branch	4	2	2
Department Store Section	15	9	6
Grocery Store Section	9	7	2
Drug Store Section	5	4	1
Vending Section	6	4	2
Product Z Sales Branch	3	1	2
Farm Sales Section	12	9	3
Building Supply Section	18	12	6
Manufacturing Supply Section	9	8	1
Mail Order Sales Section	24	12	12
	141	86	55

Figure 7-15. Sample flow chart.

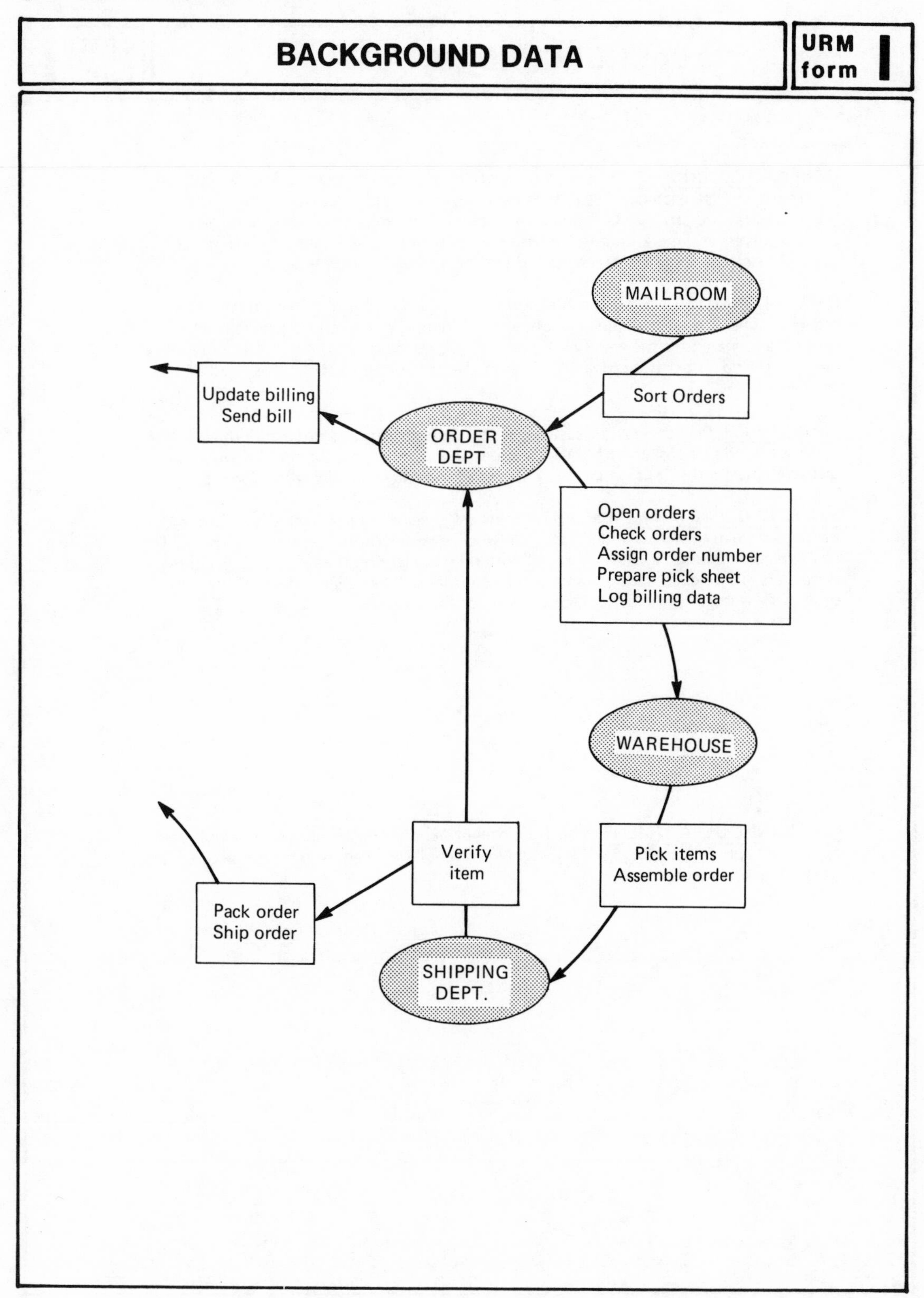

Figure 7-16. Sample descriptive paragraphs used to introduce organizational units.

BACKGROUND DATA | **URM form I**

GENERAL DESCRIPTION The company is a mail order operation. About 2000 orders are handled each day. Peak times are November and December, when 5000 orders per day are processed. Products are stocked and shipped upon receipt of orders. Buyers and marketing staff plan new items and analyze sales of current product lines. Catalogs are prepared and sent to potential customers.

ORDER DEPARTMENT Orders for products are received by this Department. The orders are checked. Customers are contacted if there appears to be a problem. Pick forms are prepared for the warehouse and billing data is set up against an order number assigned to the order. Billings are completed after the order has been packaged and shipped.

WAREHOUSE Pick sheets are received by this Department and orders are filled. Each order is placed in a basket with the pick order and sent by conveyor to the Shipping Department. Staff in this Department also stock warehouse racks with products.

SHIPPING DEPARTMENT Baskets of items are checked against pick sheets as the baskets arrive in this Department. Out-of-stock items are noted on the pick sheet. The back sheet of the pick form is packaged with items to complete each order. Shipping containers are weighed, labeled, and loaded on trucks. The original sheet of the pick order is forwarded to the Order Department for final billing.

•
•
•
•
•

(other organizational units)

•
•
•
•

PERSONNEL DEPARTMENT This Department handles recruitment of all employees, permanent and temporary. Staffing is analyzed regularly. Benefit packages are reviewed, negotiated, and managed.

Complete the Final Document (Task 4)

In this task the final document is checked, possibly edited, and assembled by the coordinator, his/her staff or publication specialists. Desired copies are made and distributed. In many cases only a few copies are needed, particularly when data are stored in computer files. The number will depend on the application for the URM data and the distribution list. Each organizational unit is normally given a copy of the entire document or, if the document is large, at least the sections where its data appear.

A large-size project does not mean that a lot of copies of user requirements need to be made. One must remember that user requirements are not usually the end. They are compared to existing buildings to define a problem or its severity or select the best solution. They are used to prepare budget requests, to evaluate design alternatives, or in many other ways. Usually a lot of people do not need a copy of the user requirements—only those involved in applying them. Some copies may be needed for historical purposes to simplify the task of keeping them up to date. URM may have been completed to serve as a political tool in getting a new lease, more space, better facilities, or a new building. The situation for which URM is used will be the main determiner of how many copies are needed.

The structure of the final document is illustrated in Figures 7-17 and 7-18. In Figure 7-17 the structure of the entire document is illustrated. Following a cover of local design and a table of contents are the background-data section, the summary-data section, and the organizational-unit-data section. If needed, additional information is placed in appendices.

In Figure 7-18, the organizational-unit-data section is expanded. The data sheets submitted by the representatives are kept in blocks as they are turned in. The sequence of organizational units is that based on the structure of the entire organization. For example, all units from a division are kept together, with branches grouped within the division. Note that the last section in the detailed data is the shared-space data. It is handled as if it were another organizational unit, having been prepared by the coordinator.

The larger an organization or a project is, the larger the resulting user requirements document will be. Summary tables and background data will be longer. There will be more organizational units and spaces. However, the structure remains the same—background data, summary data, and organizational-unit data.

The sequence of data forms within each organizational unit will remain constant, but the order of organizational units may vary for different projects. For large projects, the data will be grouped around the logical structure of an organization. For example, assume a division has two branches and each branch has two sections. The division data would be first, followed by one branch, then each of its two sections. The second branch's data would follow, followed by its two sections.

Summary

The work of developing user requirements is completed in Step 3. Representatives organize their data about their organizational unit and its spaces into final format using forms that consolidate data (Task 1). The coordinator and his/her staff use that data to prepare two parts of the user requirements document that appear at its front: summary data (Task 2) and background data (Task 3). The final document is then assembled and reproduced (Task 4) for use in some application.

After URM is Completed

After user requirements have been compiled, they will be combined with other kinds of requirements and used in some application. The role of representatives may not be finished, though. They may have to answer questions about the data of their organizational units posed by designers, staff specialists, or others involved in applying the data. Representatives may be asked to use their data in reviewing

Figure 7-17. Typical organization of components in a final URM document.

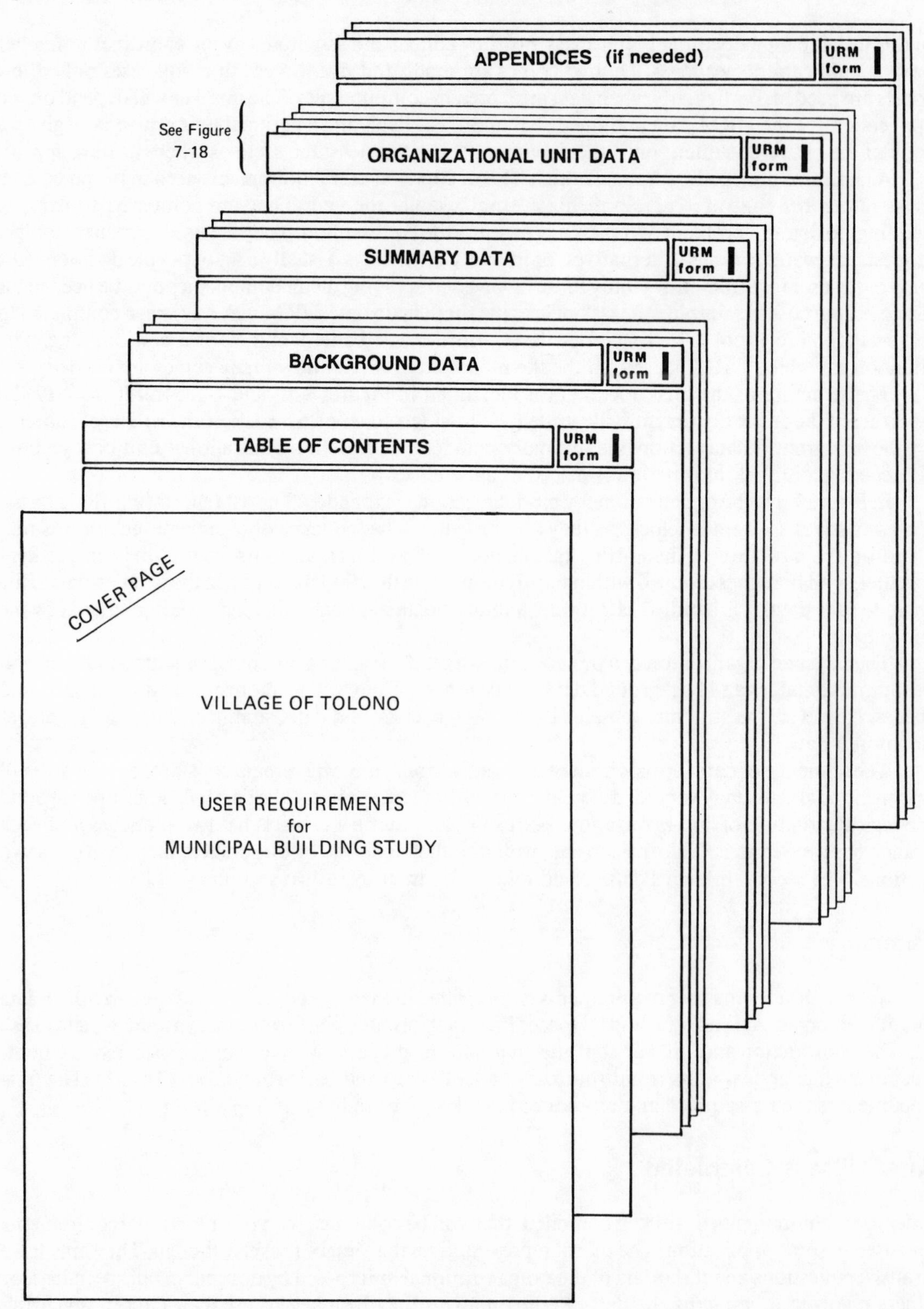

Figure 7-18. Typical organization of detailed data for organizational units in a final URM document.

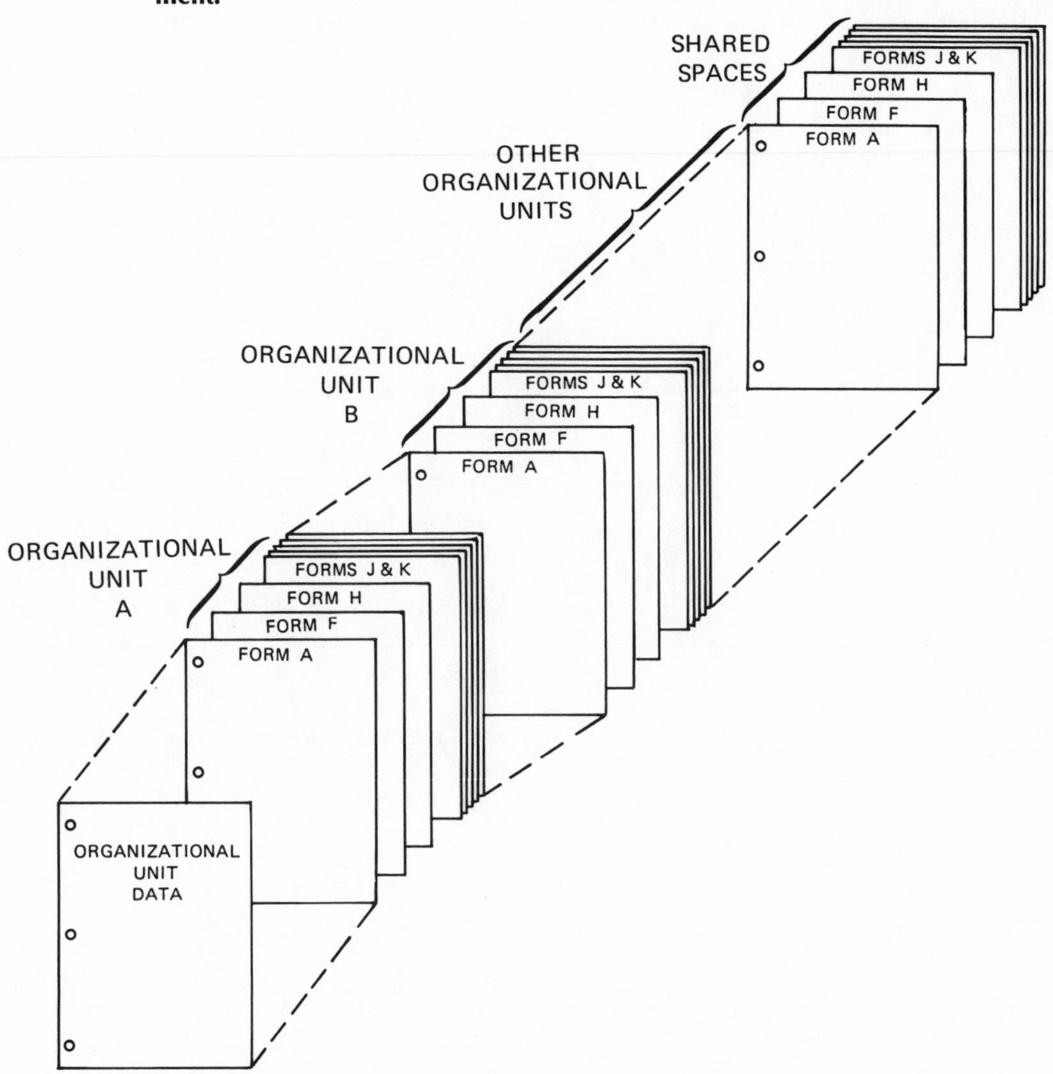

design drawings and specifications or in evaluating portions of existing buildings. Representatives may have to update data at a later date. They may be asked to define what requirements are negotiable when fitting organizational units into particular spaces and facilities. The representatives may be on call to some degree, but their involvement in the future probably will be limited.

The coordinator and possibly some staff of the coordinator may serve as the primary spokespersons for the requirements and supporting data in budget review, planning meetings, and meetings with designers and others applying the data. A designer or staff specialist who assumed the role of coordinator, would carry on after URM in subsequent roles of the coordinator. The coordinator may have to lead representatives in applying the requirements during evaluation of current or candidate buildings for lease or purchase, in reviewing designs, or in updating the data. The coordinator will likely remain as the primary liaison for the requirements for some time, perhaps for several years.

Chapter 8

URM Applications

User requirements provide the basis for many building improvement activities and the decisions made about them. User requirements are needed for early planning and budgeting, for conducting feasibility studies, for communicating with designers, for reviewing designs, for identifying problems with current buildings, and for evaluating buildings that are candidates for lease or purchase. Data from user requirements are essential for moving from one facility to another. User requirements are useful for managing incremental change and essential in facility management and space management.

The purpose of this chapter is to explore applications of user requirements and to discuss adjustments in URM that are appropriate for these applications.

Early Planning and Budgeting

After it is recognized that a building project is needed, questions about cost and schedule arise. Managers want to know what a project will cost and when it can be completed. They will want to develop a strategy for funding the project over a period of time. The two most important items of information needed in early planning and budgeting are the size of a project and the resulting cost.

Size or scope (the number of square feet or cubic feet) is one of the best predictors of the cost of a building project. There are many sources of data for estimating some measure of size. Some companies are in the business of publishing unit cost and related data for building projects. Some building and construction magazines publish cost data from recent projects or maintain indices about building costs. See the List of References near the back of the book.

Obviously, size alone will not determine the actual cost of a project. Other data must also be considered, such as the need for special features and subsystems in the building, special site preparation, adjustments or foundation problems at specific sites, the need for roads and utility extensions, geographical location, climate, and local labor costs. However, if one were to project the cost of a project from only one parameter, size would the best predictor.

Another kind of information, available from user requirements, that affects cost is the need for special features or characteristics which are not normally included in the type of building being considered.

Many examples of special features that raise the cost of a building, certain of its rooms, or subsystems could be cited. An exhaust ventilation and recovery system for a degreasing operation is a special feature for a plant. A heavily reinforced vault is an expensive special feature for a bank. Extra high and wide doors or a five-ton crane or hoist for a truck maintenance building will add to the cost of an open-bay building. Placing radiation shielding in a medical X-ray room or a radiation treatment room will cost more than normal walls, floors, and ceilings.

In early planning and budgeting estimates of size, features and cost do not have to be as accurate as later on. Initial financial considerations focus on strategies for funding and overall budget estimates. The closer the funding process moves toward authorization, the more important accuracy of cost estimates becomes. For many organizations, site selection is completed and preliminary design is initiated before project funds are given final approval. This allows specialists to develop detailed and accurate cost estimates.

As illustrated in Figure 8-1, only general requirements are needed for planning and budgeting, since they come early in the project sequence. Information that will establish the size of a project with reasonable accuracy is important. At this stage, more error in estimating size is acceptable than later on when design is initiated. It is also important to identify those features that will add to the cost of construction.

The size of a building project is based on user requirements. In URM, the total space needs for a project are developed by summing the space needs of each component of the future occupants. URM, as presented in Chapters 5 through 7, produces more data in greater detail than is needed to develop an early estimate of space needs. Since highly accurate cost estimates are not needed for early planning and budgeting, URM can be simplified for this application. Collection of unnecessary data can be avoided.

The main modification occurs in Step 2 of URM (refer to Figure 8-2). Some modifications in URM documentation, Step 3, will also occur. These modifications are detailed below. Step 1 of URM is applied essentially without change. Critical decisions must still be made about why a facility is needed (resulting in an objective statement) and who will be in the facility (resulting in a list of occupants). Occupants must still review their activities, personnel, equipment, and schedule before an estimate of size and other requirements can be made. In Figure 8-2, shading identifies which steps in URM are eliminated when the method is applied to early planning.

Figure 8-1. Only general requirements are needed for early planning and budgeting.

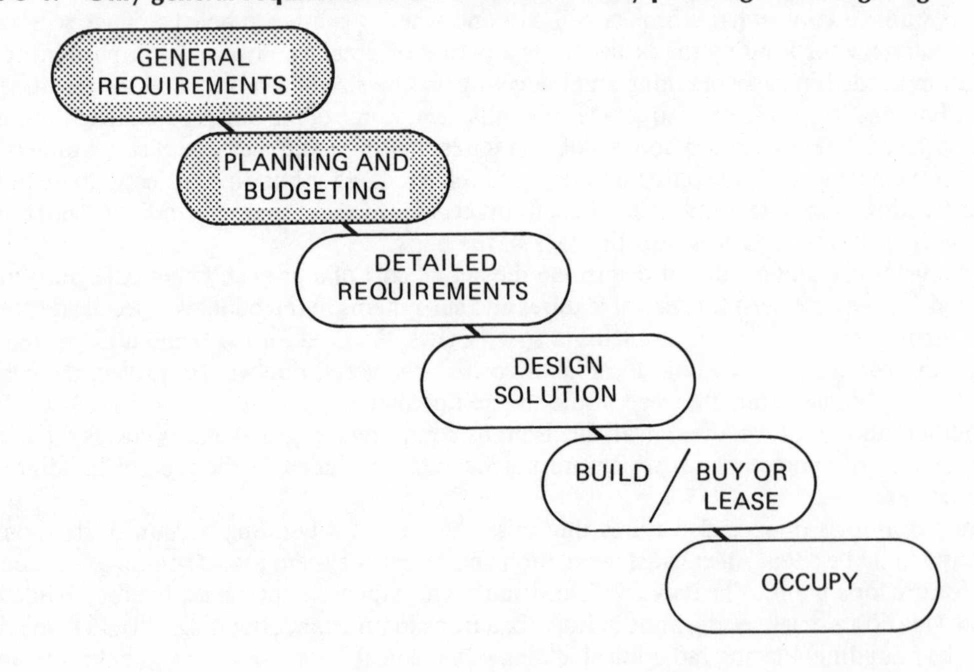

Figure 8-2. Major modifications to URM for early planning and budgeting.

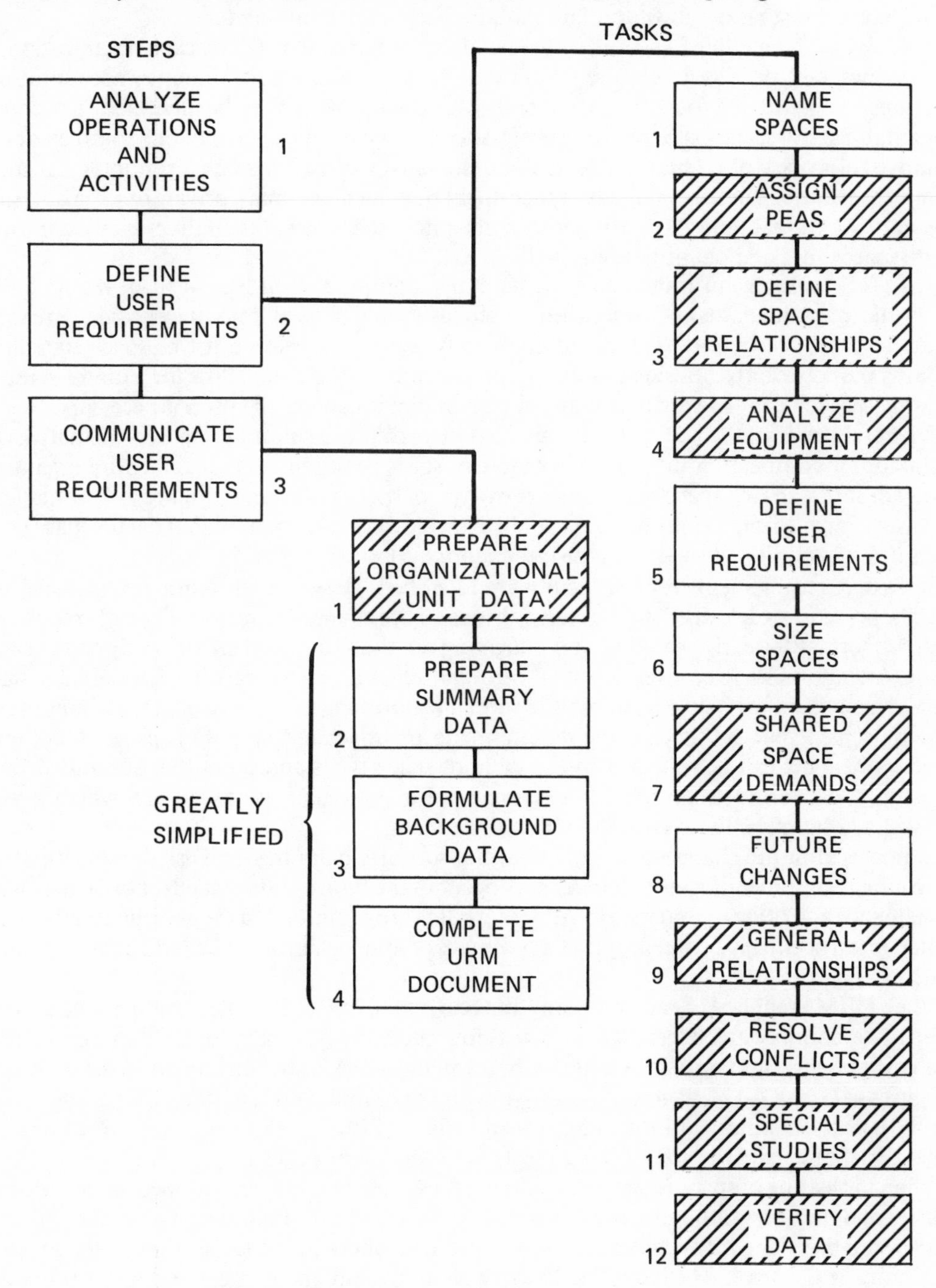

Figure 8-3 illustrates the tasks required in a modified Step 2 for applying URM to early planning and budgeting. Most tasks shown in Figure 8-2 for Step 2 are eliminated.

In Task 1 of the modified version, instead of defining particular spaces, each organizational unit merely defines the type of space required. For example, an organizational unit may need some offices, some shops, and a training room. Individual offices and shops would not be identified. The coordinator (or a staff-facility specialist or building professional assuming the coordinator's role) can develop a standard list of space types. This ensures that representatives all use the same terminology. In an office building, for example, the major kinds of space might be open office areas, private office space, support spaces and work areas, and circulation and mechanical space. Representatives can use Form I to tabulate space types and quantities required.

In Task 2, each organizational unit estimates the amount of each type of space required. Where possible, the space standards and comparison methods should be used, since they provide a reasonable estimate with the least effort. The coordinator may want to develop a list of space standards for estimating purposes and distribute that list to representatives. This will allow for a uniform means of estimating space needs. The analytical and sketch methods can also be used if necessary.

The results of Task 1 and 2 can be organized in a table and submitted by representatives to the coordinator for compilation in a summary table of space requirements for the entire organization. One way of summarizing space requirements by type of space is shown in Figure 8-4. Other formats can be used. For example, it may be appropriate to create a cross-tabulation that details space demand by organizational unit and by space type (see Figure 8-5).

In Task 3, requirements for the facility are identified. However, all requirements need not be listed. The procedures described in Chapters 5 through 7 are changed here, too. The only requirements that are important for early planning and budgeting are those that add significantly to the cost of a building. Attention should be given only to those requirements that will result in special subsystems or unusual demands on the building, or require special features that are not typical for a building of the type being considered. In this task, the requirements are attached to types of spaces. A column for costly requirements can be included in the table of space types and quantities submitted by each representative on Form I, as illustrated in Figure 8-4. If representatives are not sure whether requirements will add significantly to cost, they should list them.

Estimates of future changes also are important for early planning. In Task 4, representatives of organizational units identify what changes may occur in the future and what effect these may have on space demands and other requirements. After these data are submitted to the coordinator, they can be tabulated for all occupants. An example of a summary of future changes and effects on requirements is provided in Figure 8-6.

When URM is applied to early planning and budgeting, URM documentation is greatly reduced because there is simply much less data. As few as four pages are necessary for the final compilation of data: a cover page, a summary sheet that combines an objective for the facility and a list of occupants (see Figures 4-11 and 4-12), a table of space types and quantities and requirements for each (Figures 8-4 or 8-5), and a summary of future changes and effects (Figure 8-6). Depending on the number of organizational units and the size of the project, more than four pages may be needed.

When the leasing of space is being considered, these data also may be grouped so that acceptable floor sizes and numbers of floors can be established for multistory buildings. General requirements compiled into these and other tables are used to evaluate alternate buildings for general fit.

In summary, when URM is applied to early planning and budgeting, the most important data from URM are the overall size of the building project and special requirements that might raise the cost of the project above that for a normal building. Step 1 is implemented as initially described in Chapter 5. Steps 2 and 3 are modified because much of the detailed data important for communicating with designers is not needed in this stage of a project.

Figure 8-3. Step 2 of URM when applied to early planning and budgeting.

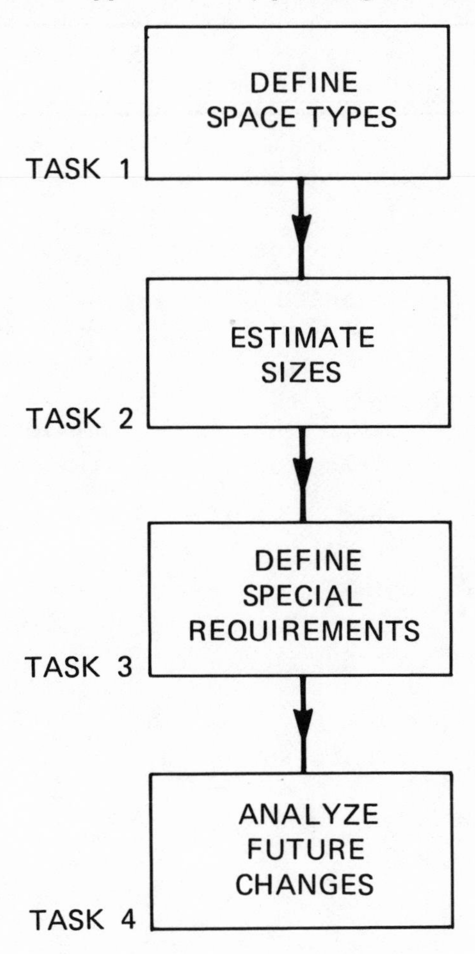

Figure 8-4. Summary of space needs by type of space.

URM form I

TYPE OF SPACE	QUANTITY REQUIRED (SQ. FT.)	MAJOR REQUIREMENTS
Offices - private	800	Luxury finishes
Offices - open plan	2000	
Conference room	400	Sound system
Shops	3550	Compressed air, 220v-3 phase, exhaust system
Truck bays	6500	Heat, water
Restrooms	450	
Storage - general	2500	
Subtotal (net space)	16200	
Circulation & mechanical (20%)	3240	
TOTAL	19440	

Figure 8-5. Summary of space needs by type of space and organizational unit.

URM form I

TYPE OF SPACE/ORGANIZATION	QUANTITY REQUIRED (SQ. FT.)	
Offices - Private		
Top Management	600	
Division A	200	
Division B	200	
Division C	200	
Division D	200	
Subtotal		1400
Offices-Open Plan		
Top Management	200	
Division A	1250	
Division B	1800	
Division C	950	
Division D	2375	
Subtotal		6575
Shops		
Division C		1575
Restrooms		1250
Storage - general		3500
Total (net useable space)		14300
Circulation & Mechanical (20%)		2860
TOTAL		17160

Figure 8-6. Summary of future changes and impacts.

URM form

ORGANIZATIONAL UNIT	NATURE OF CHANGE	WHEN EXPECTED	IMPACT ON REQUIREMENTS
Police Dept.	New radio system	1 yr	50 sq. ft. additional 30 ft. antenna tower
	Add third shift	2-3 yrs	100 sq. ft. office
Public Works Dept.	New water testing process	3 yrs	200 sq. ft. lab Environmental controls
Village Office	Microfilm equipment: camera, reader	6 mo	80 sq. ft. for equipment
	Microfilm library		50 sq. ft.

Feasibility Studies

Feasibility studies for building projects can have a variety of meanings. A feasibility study may seek to compare financing strategies or determine if an organization can afford to implement a project. It may be used to develop and evaluate general concepts for a solution. These concepts may apply to an existing building and site to see if expansion is possible or to new, undeveloped sites. A feasibility study may be used to compare a variety of sites with the goal of selecting the best location. It may determine if a project will be in compliance with local regulations and ordinances and if local officials will consider zoning changes, financial incentives, or other actions that will influence cost and site selection. A feasibility study may compare availability and capacity of utilities (sewer, water, electricity, gas), roads and traffic, or economic and other impacts on local communities or neighbors. It may compare leasing, buying, new construction, and remodeling. In any of these cases, the objective of a feasibility study is to determine if a proposed building project is financially, physically, and legally possible.

Data from URM are essential for feasibility studies. In most cases, the data developed for early planning and budgeting (as described above) are all that are required. If design concepts are to be compared or evaluated, more data will be needed. But all the details resulting from URM as described in Chapters 5 through 7 will probably not be needed. The detail required will be somewhere between that needed for planning and budgeting and that needed for full design.

The kinds of space and amounts required must be established for each occupant organization. The overall relationships among organizational units and major spaces or kinds of space must be identified. More information about requirements may be needed to estimate power, fuel, sewer, and water demands. The objective for a feasibility study will determine what URM data is needed.

The procedures recommended for compiling planning and budgeting data can be used for feasibility studies. If more details are needed, portions of the URM procedures defined in Chapters 5

Figure 8-7. Detailed requirements are needed for design.

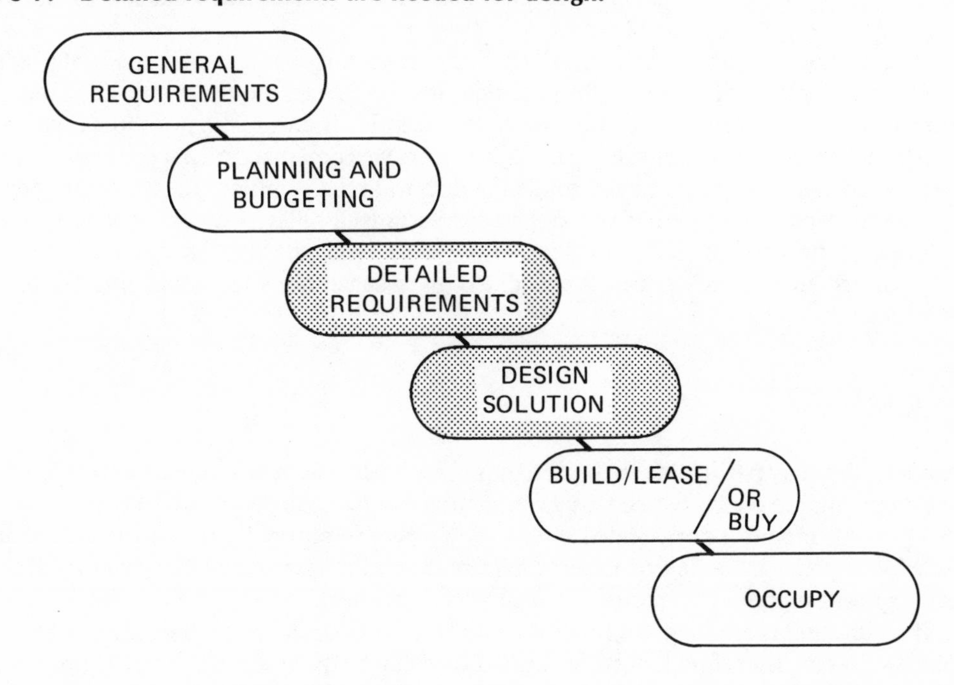

through 7 can be added. Because the term *feasibility study* has a variety of meanings, no firm rule for adjusting URM can be stated. If it is clear that a project will move forward and a feasibility study will focus mainly on site selection and financial and legal matters, it may be best to complete full URM right away. Then the details for a feasibility study will be available and need only be extracted from URM documentation. The URM data will be used again as the project moves to design.

Design

The design for a project (new construction or renovation) cannot be started unless user requirements are available. If they have not been developed by the time a design firm has been selected and project design begins, then the designer will ask questions intended to establish user requirements. See Figure 8-7.

The design process begins with a general solution and progresses toward a detailed one. At first, summary data is important. Later, details are needed. At the beginning of the design process, an architect is interested in large blocks of space, organized by space type and organizational unit. After space is organized to establish the building envelope or shape (a process called *massing*), detailed features and characteristics are worked on. Later, the layout and organization of individual spaces, and even work stations, will be addressed.

As noted at the beginning of this book, a major goal of URM is to provide an orderly means by which users can communicate what is needed to designers. The documentation of user requirements described in Step 3 of URM anticipates the information needs of the design process. Both summary data and details are provided. The format in the documentation allows the designer to locate information easily during design. As illustrated in Figure 8-8, a hierarchy of data forms the basis for the format.

The background data introduce a designer to the organization, its units and operations, and why URM or a project was started.

The summary data give an overall feel for the size of the project and what spaces make up the project.

The data for each organizational unit tell the designer what it does and what spaces it needs. Data for each space are organized into a two-page format that gives size, requirements, and contents.

Organizing URM data into a document is very helpful. It ensures that the data will be used, because a designer can study it easily. Even though data are well organized on paper, they are presented in only one sequence. One must page through the data to find all similar spaces, requirements, and activities—a time-consuming job. Even if they were arranged differently, one would have to page through them to find similar items of data. However, having the data in computer files makes searching through them much easier. Appendix B gives suggestions for organizing URM data in computer files.

Design Review

Users can meaningfully participate in the design process when user requirements are clearly defined. When requirements are poorly defined, users frequently assume requirements that may or may not be correct. They will modify poorly stated ones to their own interpretations. This results in useless, misdirected, and irrelevant review comments. The main role for users during design is to evaluate how well the design will support their requirements. Users must check design drawings and documents to find out if requirements are met. No modifications to URM or additional data are needed for user design review. In fact, users find that reviewing a design can be quite simple if their requirements are well defined.

Figure 8-8. Hierarchical structure of URM data.

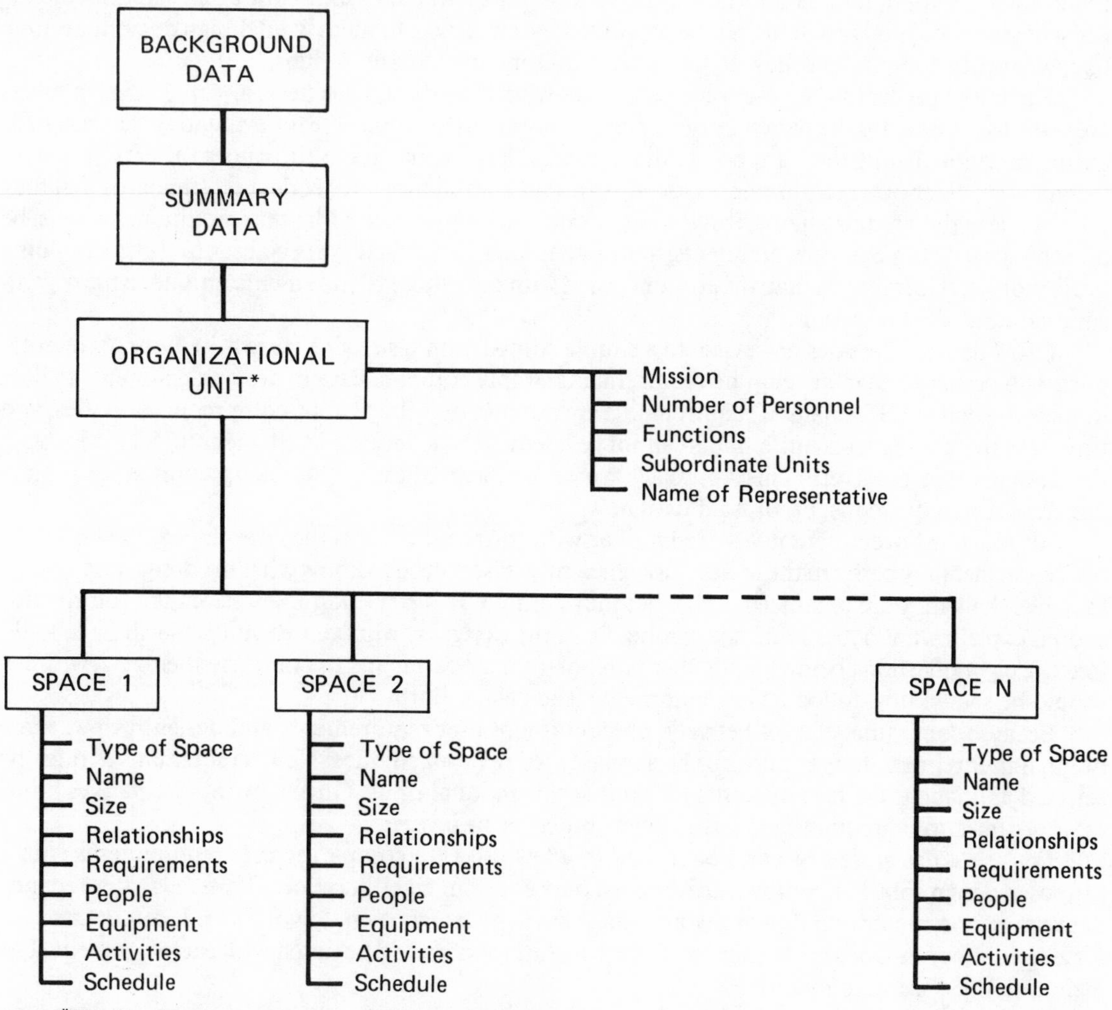

*Similar data structure for each organizational unit.

User participation in design review is not without problems, even when users understand their requirements. Often, they don't know what to look for or where to look among design documents. Most of the procedural problems can be prevented by explaining to them what design documents look like, what they contain, and how to compare requirements to data in them.

The URM process shows user representatives what they should be checking for. Representatives are responsible only for the data they developed. They are not responsible for evaluating the design for an organizational unit they are not familiar with. They do not have to concern themselves with technical, legal, or other requirements, or worry about compliance with codes and standards that they did not cite in their requirements. Reviewing a design for compliance with other requirements must be done by specialists (engineers, architects, interior designers, staff-facility specialists, and other building professionals). User representatives are concerned with only those requirements unique to the organizational units they represent.

URM data reduces design review to a simple comparison task (see Figure 8-9). User representatives who participated in developing requirements simply compare data in design documents to data in their respective URM data. If requirements are not met or if the people, equipment, activities, and time data that drive user requirements cannot be adequately accommodated, meaningful feedback to the designer can be given. This approach works well regardless of the completion stage for the design—schematic, concept, or final design.

Although user representatives are familiar with the requirements they developed, they must be trained so that they perform the review task efficiently. Users need to know what the design documents look like, how they are organized, and what they contain. In a brief training session, the coordinator should explain what documents are coming from the designer, what representatives will have to do (see the eight questions below), and where to look in the documents to complete their review. They should be shown how to log review comments (such as on Form L).

Because some time elapses between preparation of user requirements and design review, some original user representatives may not be available for review activities. New representatives must be selected and taught the requirements for their organizational units. Once new representatives know their organizational requirements, the review process can proceed.

For users, design review can be reduced to a few tasks. In comparing user requirements with a proposed design solution, representatives need only consider eight questions. If there is a discrepancy between requirements and design documents, comments are written down. Form L can be used for logging comments. Document, page, and drawing numbers should be noted with each comment. The eight questions are as follows:

1. *Are the spaces you need included in the design?* Reviewers should look through the floor-plan drawings to find every space included in their user requirements. Discrepancies should be noted.

2. *Is each space the correct size?* The amount of space listed in user requirements should be compared to the amount provided in the drawings and in any table of spaces and sizes. This may involve measuring the scaled drawing and computing the resulting size in the design. There are bound to be some differences between the amount of space listed in user requirements and that found in a design. A reasonable discrepancy should be accepted, because a designer may be confined to certain increments (a four- or five-foot grid system is commonly used) of length. Users must consider discrepancies particularly in terms of the effects on activities and equipment.

3. *Are spaces organized correctly?* Adjacency matrices and bubble diagrams included among user requirements should be compared to the layouts provided in the design. Spaces that must be next to each other should be indicated on drawings. Spaces which can be or should be farther apart must be checked for acceptable distances. Overall relationships among organizational units and major- or shared-use spaces should be analyzed to see if overall arrangements are suitable. Site plans must be consulted in addition to floor plans, especially when exterior spaces and exterior circulation are involved in relationships. Unacceptable relationships should be discussed.

Figure 8-9. User requirements are compared to design documents in design review.

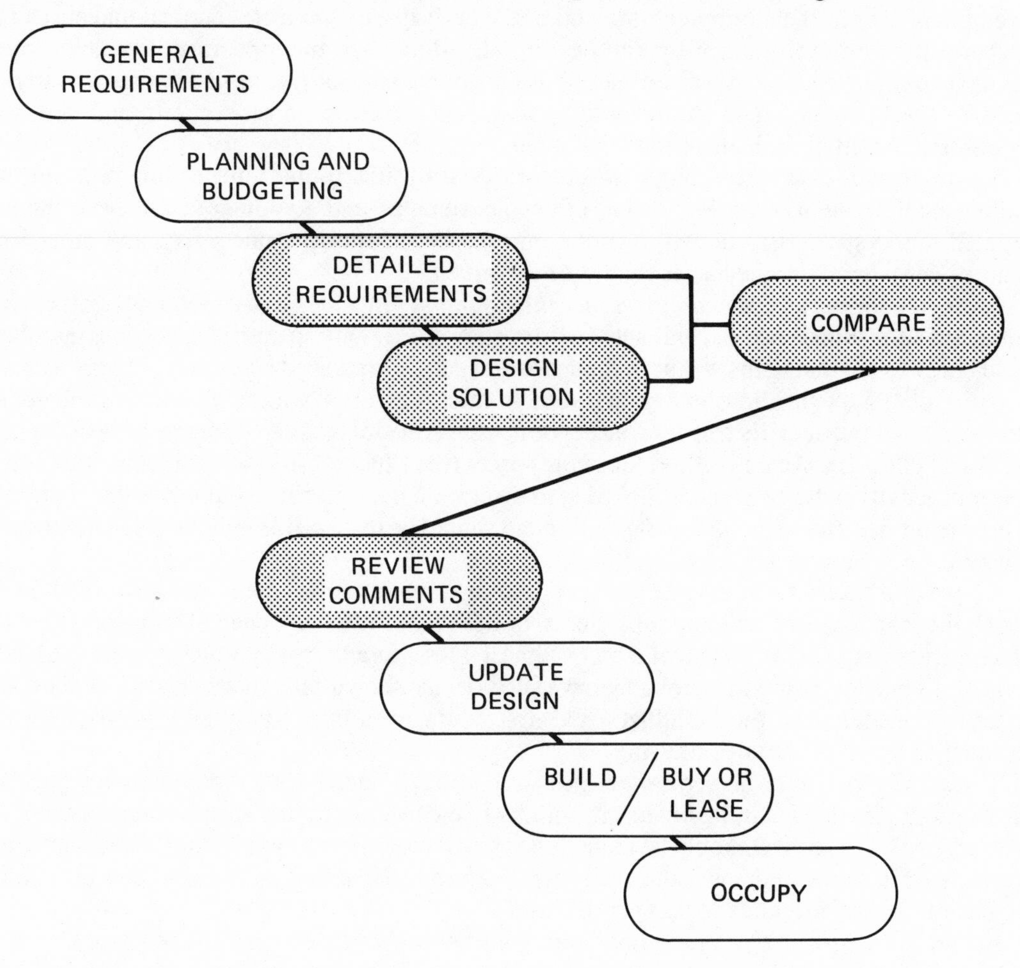

4. *Is access or circulation adequate for each space?* Representatives must check to see that movement of people and equipment into and out of each space is possible. They should check to see that items to be moved into a space can pass through doors and that operational vehicles (such as forklifts) can get in and out. Special dimensions for doors and windows logged in user requirements should be checked. Data here are likely to be found on elevations, cross-sections, and plan views. Obvious and potential problems should be noted.

5. *Are general circulation routes satisfactory?* Main traffic routes into, within, and out of the building should be looked at. Movement of people, vehicles, and equipment on the site should be evaluated. Congestion, peak activity periods, emergency situations, traffic safety, and other factors should be considered. Inadequacies should be explained.

6. *Are the shapes of spaces or layouts for them suitable?* Included in user requirements is a list of equipment, furnishings, people, and activities for each space. During design review, representatives should check to see that items will fit in the space and can be organized effectively. In some cases, the shape of a room may not allow for a good layout. In many designs, layouts within spaces will not even be shown. This is particularly true for concept designs. A scaled sketch can be made to check for layout problems. Follow the sketch method for sizing spaces (see Chapter 6). Door placement and internal traffic and activity patterns must be looked at to determine if a suitable layout is possible. Extra areas required to service, maintain, or handle equipment should be reviewed as well. Deficiencies should be discussed.

7. *Are other user requirements satisfied and features provided?* Representatives should look through the schematics of building subsystems, or floor plans with subsystem details shown, to see if such requirements as lights, electrical power, exhaust lines, lifts and cranes, water, or other utilities are provided. Concept designs will probably have such details shown only in schematics of subsystems that are to be built into the building. Final design drawings may have many of these elements illustrated or noted on detailed drawings.

8. *Can user activities be performed effectively and efficiently?* User representatives must keep their missions, functions, and activities in mind while checking on the specific items listed above. However, one last overview should be made. The overall building concept, layout, and design should be considered in terms of the way the users must perform their activities. A macroview may turn up deficiencies not noticed while looking at details.

Just as in documenting user requirements, comments about a design must be stated clearly and presented in an organized manner if they are to be understood and acted on by the designer. Comments should identify what is wrong with the design and, where possible, offer suggestions for a solution. Comments that merely criticize the design or refer to things about which the designer cannot do anything are of little help. Comments should refer to specific drawings or pages in particular design documents and be organized in some logical manner. One may refer to user requirements if they have not been met. One format for logging user design review comments is illustrated in Figure 8-10.

Evaluating Existing Buildings (Candidates for Lease or Purchase)

URM provides valuable data for evaluating existing buildings that are candidates for lease or purchase.

Feasibility studies have been discussed primarily in terms of new construction. If lease or purchase is being considered, feasibility studies might include evaluations of existing buildings. URM data will provide the basis for determining whether an existing building can meet functional needs. Cost, legal, and regulatory constraints and technical requirements are also important when an existing building is being considered. These factors must be handled by building professionals in evaluating a design.

Figure 8-10. Example of user design review comments entered on Form L.

REVIEW COMMENTS

URM form **L**

reviewer		organizational unit	date
Reference: drawing/page/ space name or number	**comment number**	**comment**	
Drawing P - 14, Space 8 -3	1	The Fire Chief's office is not shown in the floor plan. It appears to be missing.	
Drawing P - 16, Space 2 - 5	2	The office for the Public Works Director is about 20% smaller than required.	
Drawing P - 16 Space 2 - 4 and 2 - 5	3	The drafting room is not directly accessible from the Public Work Director's office. It must be adjacent with a connecting door.	
Drawing P - 22 Space 7 - 1	4	It appears that there is insufficient turning space for a forklift in the Storage Bay/ Entrance area. The Village uses an oversized forklift.	

Figure 8-11. General feasibility of lease or purchase alternatives can be determined by comparing general requirements to each alternative.

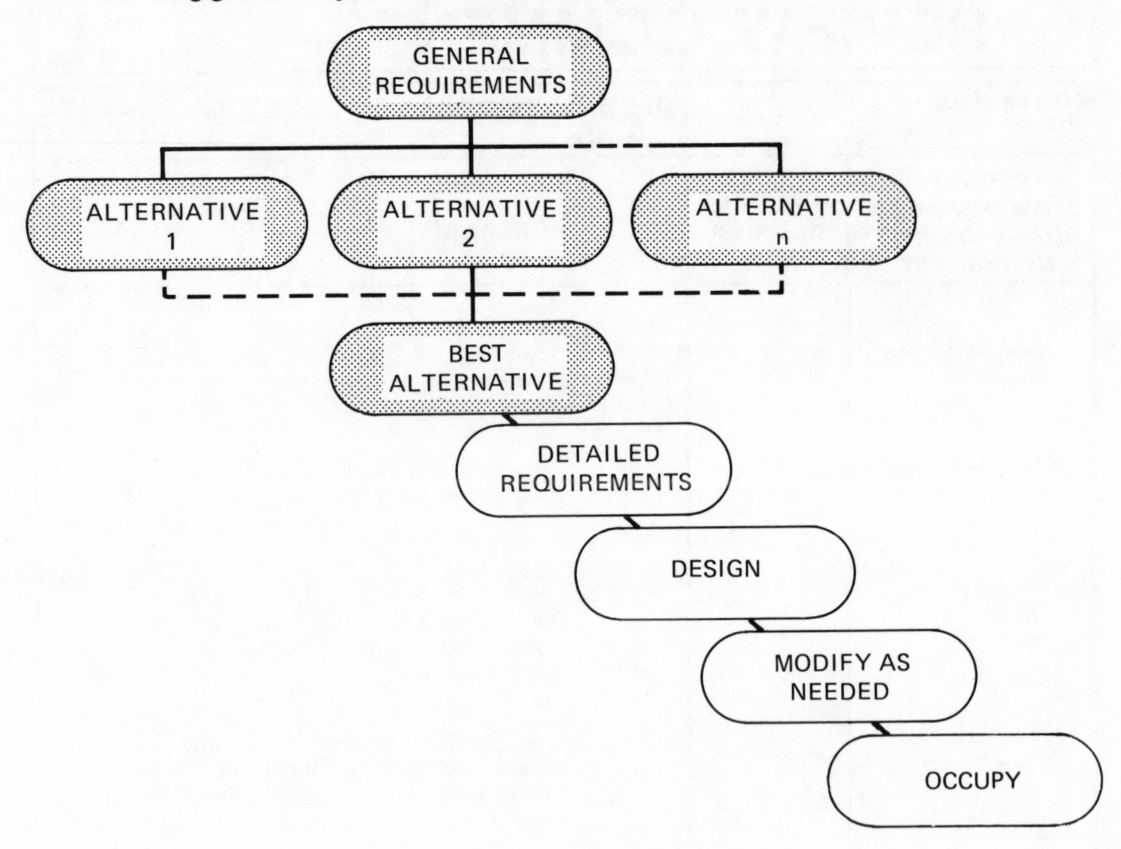

Figure 8-12. Final selection of lease or purchase alternatives can be made by comparing detailed requirements to each alternative.

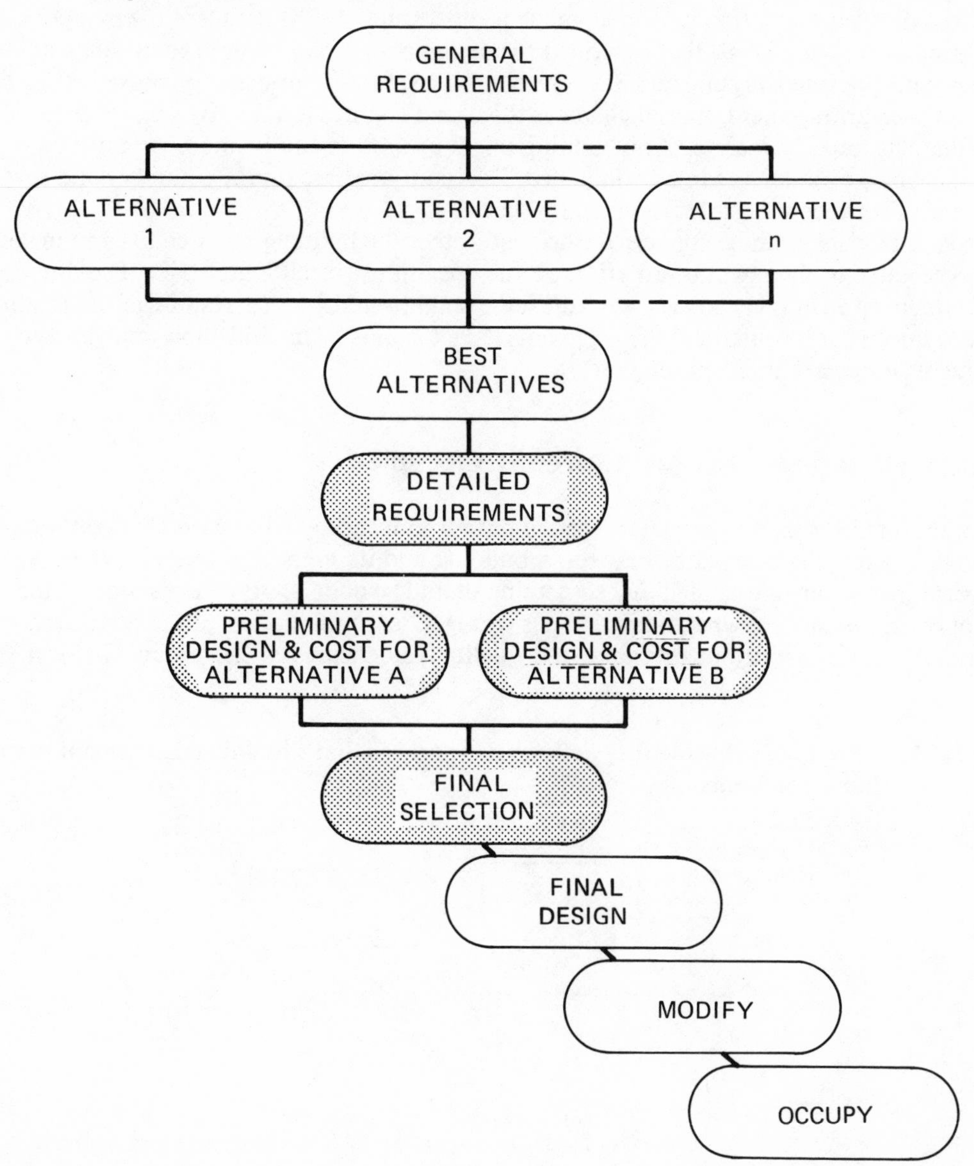

If one is concerned with general feasibility, general requirements, as developed in the planning and budgeting modification to URM, are all that are necessary (see Figure 8-11). If a final selection is to be made, a detailed evaluation should be made using detailed URM data (see Figure 8-12).

A process very similar to that described for design review can be used to evaluate an existing building, both quantitatively and qualitatively. Kinds and amounts of space (quantitative) are considered first. Then arrangement, special characteristics, and features (qualitative) are reviewed. Often a design that illustrates how an existing building might be used or can be modified must be prepared before a detailed evaluation can be conducted. Decisions are based on the number and severity of discrepancies between what is required and what exists.

Each discrepancy can be converted into cost. Either the building must be fixed to make occupancy acceptable or there will be an effect on the operations (recall Figure 1-2). The discrepancies should be converted into 1) a total cost to adjust the building and 2) a cost resulting from productivity being less than what it could be if the occupants have to adjust. This will allow analysts to compare alternate buildings in a uniform manner.

Evaluating Buildings That Are Currently Occupied

A major use for URM data is in evaluating buildings currently occupied by users. As organizations and operations change, the original fit between a building and its users deteriorates. When operations deteriorate greatly, anyone can tell that something should be done. Before things become that bad, it may not be easy to observe what, if anything, is wrong. User requirements can be compared with the current building to identify quantitative and qualitative deficiencies (see Figure 8-13). If there is

Figure 8-13. URM data helps identify deficiencies early so that a building can remain supportive to its occupants.

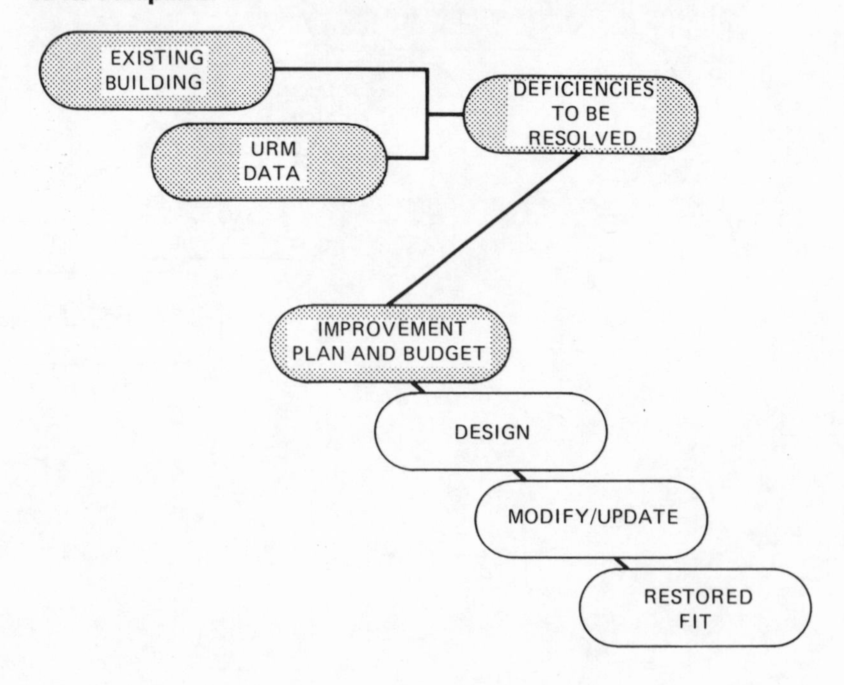

Figure 8-14. Example of a table in which existing spaces are compared to required ones to identify deficiencies.

	QUANTITY			QUALITY
ORGANIZATION/SPACE	Required	Existing	Difference	Deficiencies
VILLAGE OFFICE				
Visitor Area	60	40	-20	No counter
Main Office	200	130	-70	Poor lighting for one desk
Village President's Office	120	150	+30	Tear in carpet, soiled walls
Store Room	250	125	-125	Insufficient shelves
Records Vault	80	0	-80	
Meeting Room	400	410	+10	
Total	1110	855	-255	

competition for improvement funds, precise definitions of deficiencies can provide the justification to beat others to the available improvement funds.

The procedure described for design review above can be used to evaluate an existing building as well. Requirements are compared to the way things are and deficiencies are noted. Again, Form L can be used to log evaluation results. Quantitative results can be compiled in a table of excesses and deficiencies by organizational unit or type of space. An example is shown in Figure 8-14. Results can be used to seek additional space, dispose of excess space, or redistribute space.

Qualitative deficiencies can be similarly tabulated from comment sheets. Like kinds of deficiencies (paint, lighting, flooring, et cetera) can be grouped so that work orders can be prepared or contracts for repair, maintenance, or improvement formulated and assigned.

URM data must be kept up to date if it is to keep buildings supportive of users. It also requires that a review of an existing building be completed periodically. How often the review should be done depends on the rate of change experienced by an organization. For many organizations, a review every one or two years should be sufficient. For rapidly changing organizations, a review every six months will be needed. For organizations that change very slowly, a five-year cycle may be appropriate.

Prior to conducting the evaluation, the last available set of URM data must be updated. New organizational units will have to develop their requirements. Existing units will merely have to update their data.

Moving

In order to plan for a move, one needs to know what is to be moved, where it is, and where it is to go. Then a schedule can be worked out to minimize interruptions and moving costs. URM data contains much of the information necessary for planning a move. The locations where things are to go (space names) and the items to be located in each space are found in user requirements data. Only current locations of items and scheduling of the move need to be added. URM data may have to be merged with inventory data to provide a complete list of things to be moved.

If new equipment and furniture is to be purchased, URM data can be used to compile the amounts that are required. If existing items are to be mixed with new items, it is easy to mark the URM data accordingly.

While URM data about equipment and furnishings may not be complete, they provide the basis for preparing for a move and planning new equipment purchases related to a move.

Standardized Spaces and Facilities

Some facilities and certain kinds of spaces lend themselves to standardization. As noted earlier, standards are frequently looked at as allowances. Offices, schools, living accommodations (such as dormitories, jail cells, apartment units), repair shops, fast food restaurants, and theaters are some examples of facilities for which user requirements (and design criteria) can be standardized.

It is easier to standardize data for individual spaces within a facility than for a whole building. Assumptions must be stated about people, equipment, activities, and scheduling (PEAS) for which standards are created and typical user requirements (and related design criteria) can be compiled in a reference document. The assumptions are essential to assess the suitability of standards to a particular application. Care must be taken in applying standards to ensure that the assumed PEAS are identical to those of the real occupants. If assumed PEAS are not identical, the standard will probably not be applicable entirely and user requirements will probably need to be adjusted for the real occupants. An example of standardized URM data for one space type is provided in Figure 8-15.

When spaces and facilities are standardized, the URM process can be changed to simplify it. Instead of estimating space size or listing requirements, the appropriate space standard can be referenced. When the standard does not completely accommodate users, exceptions only need to be logged in URM for each space. Use of standardized user requirements data can simplify URM considerably and reduce the time required to implement Steps 2 and 3 (see Chapters 6 and 7).

Facility and Space Management

In the late 1970s, the term *facility management* became popular. There is no precise meaning for the term. In general, facility management recognizes that buildings consume a large amount of capital, but by themselves produce nothing. Facility management is an orderly process for managing the incremental changes needed in buildings that result from changes in organizations, their size, activities, and supporting equipment and staff. Many companies and organizations have shifted from a caretaker philosophy to a management philosophy. Facilities are to be managed like any other resource.

There is no standard method for facility management. The approaches are almost as varied as the organizations that have implemented facility management. Here's a simplified, three-step description of facility management activities: The first step is identifying problems with facilities. The second is organizing problems and prioritizing solutions in an improvement plan. The third step is implementing the plan.

Space management is closely related to facility management. Space management is concerned with distribution and utilization of current spaces and projection of future space needs. Space management may be an element of a facility management organization or be a separate function. The main differences are scale and focus. Facility management tends to deal with large elements of space (whole buildings or major parts), while space management addresses types of space and individual spaces or portions of buildings. Both involve planning and identifying problems with existing buildings or spaces and resolving deficiencies.

URM is an important tool for facility and space management. In order to determine if something is wrong with an existing building, one must know how things should be. URM data states explicitly what spaces and characteristics are needed. Deficiencies are identified when an existing building and its characteristics are compared to URM. As organizations change, user requirements data can be updated and a regularly scheduled evaluation of existing facilities can be conducted. The facility managers will use results to formulate plans and implement improvement actions. The space managers can use the data to dispose of unneeded space, consolidate activities, reallocate space, resolve shortages, or project additional space needs.

User Requirements for Ill-Defined Organizations and Operations

Sometimes in planning a facility it is not clear how an organization will be structured or how operations will work. It is difficult to define requirements and plan a building when the organizations that will go in it do not exist. It is also difficult when it is known that organizations in the building will be changed, but the change has not been finalized. It is also difficult to plan when it is known that operations will be modified, but the modifications haven't been resolved. One can put up a building in spite of these information gaps. However, it may not provide a good fit for occupants when they move in.

For example, a company may have decided that it will operate its training program at one central location instead of using several remote locations. New instructional programs will be incorporated,

Figure 8-15. Example of standardized requirements for classrooms.

Standard Requirements for Classrooms

Use/Activity	Classrooms are typically used by one or more instructors to conduct lectures, presentations, or demonstrations, using a variety of training aids. The primary activities of students in the classroom are seeing, hearing, and writing.
Occupants	The number of instructors, including teachers' aides or technicians, may vary from one to four. The size of the audience will range from 50 to 100 students.
Equipment/Supplies	The instructor needs a platform, chalkboards, tackboards, map hangers, projection screens, and equipment for demonstrations at the front of the room. A lectern, table or desk may also be needed. Desks may have to be arranged in temporary or permanent tiers to enable students to see the instructor and/or training aids. These desks should have a writing surface. Tables and chairs may also be used. Projection or sound equipment that is kept permanently in the classroom should be placed on movable stands or mounted securely. Other demonstration/training-aid equipment can be kept in a storage area adjacent to the classroom.
Space/Size	Classroom space should be sized to support a variety of classes, instruction methods, and classroom activities. There must be enough space near the front of the classroom for audiovisual and other teaching equipment.
Space/Shape	Classrooms should accommodate a variety of activities and layouts. Seating arrangements should provide good eye contact between the instructor and students. Ceilings should be high enough to accommodate projection screens.
Access/Circulation	A classroom should be conveniently located in the building and away from noisy areas. Late students should be able to enter the classrooms without disrupting the class. Circulation routes should be provided around the seating area. Movement of equipment in and out of classrooms should occur with ease.
Utilities/Waste	Electrical service should be capable of supporting a variety of demonstration and teaching equipment. Lighting controls should be located where the instructor can easily reach them.

Figure 8-15. Continued

Environmental Conditions	Various lighting levels are needed to meet lighting requirements or any type of instruction.
	Daylight entering the room through windows must be controlled to minimize shadows and glare.
	The instructor should be easily heard.
	Sounds from outside the classroom should not enter the room and disrupt the classroom activities.
	Thermal conditions should be comfortable.
Appearance/ Finishes/Image	Floors should be attractive, easy to maintain and functional.
	Interior finishes and colors should be attractive.
Communication	Students at all locations in the room should be able to hear the instructor.
	There should be enough television monitors to insure that each student can see one well.
	The instructor should have easy access to sound controls.
Storage	General storage should be provided in each classroom for frequently used equipment.

but they haven't been defined. Teaching methods, staff size, and structure are also unresolved. A major decision also remains: Will courses be primarily self-paced, using computer-aided instruction, or will traditional 20-to-40-person classes be conducted using a standard batch mode for student progress through the program?

URM can be helpful in cases where operational and organizational options have not been resolved. URM helps planners think through the facts of the options and the implications for the facilities. If URM is completed for each of the options, results will help establish cost and effectiveness for each option (see Figure 8-16).

The early planning and design approach is probably sufficient to help establish project size and cost for each option.

In the training facility example above, a self-paced instruction method will result in entirely

Figure 8-16. URM can help in the selection of options for organizational structure or operation methods.

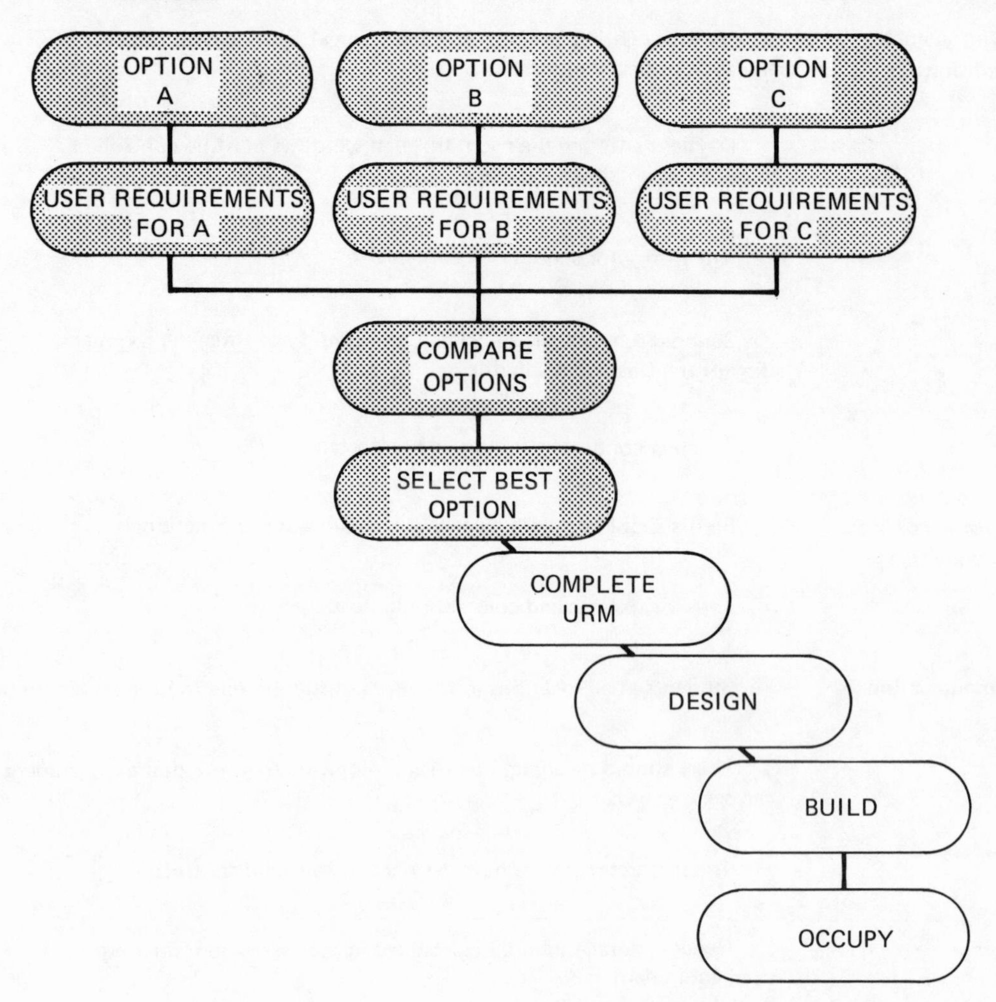

different kinds of spaces, sizes, and features than will the classroom training method. By developing user requirements for each training method, one can compare equipment and facility costs for each method against the cost of training per student. The cost of equipment, staff, and buildings relative to the value of student training time will reveal which method and facility is the better investment.

When a method of operation or organizational structure is finally selected, preliminary URM data can be expanded for the selected option. Detailed URM data for designers can be prepared by updating the original, using the procedures in Chapters 5 through 7.

If one is considering buying or leasing a building instead of new construction, the three-step process described above can be combined with building evaluation and selection procedures illustrated in Figures 8-11 and 8-12. This will allow one to apply URM to multiple operational and organizational options while facing multiple building alternatives.

Summary

URM has many applications. URM data defines how things should be. In many of its applications the way things should be in facilities are compared to the way they are or might be. URM data can be compared with currently used buildings as a means for defining facility problems. It is also possible to compare URM data with existing buildings that are candidates for leasing or purchase. Quantitative and qualitative comparisons will identify how well they might support an organization. URM data can be compared with designs to determine if they will provide a good fit for the future occupants.

URM procedures and data can be adjusted to fit the level of accuracy and information detail needed for different kinds of facility decisions. Simplified URM steps and data can provide reasonable information for early planning and budgeting and for feasibility studies. Detailed data can be generated by URM for final budgets, getting designs underway, making lease and purchase decisions, and move-in preparation.

URM can be used for nearly any kind of project. It may need to be supplemented for manufacturing and other equipment-intensive facilities. It can be applied to small, medium, and large projects and to simple or complex facilities.

URM can be applied to large and small organization facility problems. URM can be applied by any organization: private, public, governmental, business, religious, educational, medical, industrial, or institutional.

URM can be applied by guiding professionals (engineers, architects, and other designers), facility specialists within organizations, future occupants themselves, or facility committees.

URM can be used to define requirements of existing organizations, of organizations being formulated, or of organizations undergoing change.

URM can be applied for establishing facility standards.

URM is a flexible, simply structured, logical, and orderly procedure for defining how facilities should be.

List of References

Architectural Programming

These books are written primarily for architects and related design professionals. *Architectural programming* is the process of preparing technical, functional, and budgetary data for facility projects before the design for a solution is initiated. URM is a major portion of the programming process. Most of these references do not provide a detailed, step-by-step process. Some provide ideas for dealing with special situations. Others provide details about technical data for building projects (see Appendix D).

Gould, Bryant Putnam. *Planning the New Corporate Headquarters.* New York: John Wiley & Sons, 1983.

(This book applies to administrative office buildings only.)

Kemper, Alfred M. *Architectural Handbook.* New York: 1979. John Wiley & Sons, 1979.

(One section of this handbook is devoted to architectural programming. There is limited information on developing user requirements. Most information is devoted to technical and budgetary matters.)

Palmer, Mickey A. *The Architect's Guide to Facility Programming.* New York: AIA and McGraw-Hill (Architectural Record Books), 1981.

(This is the most complete technical reference available on architectural programming.)

Peña, William M., William Caudill and John Focke. *Problem Seeking: An Architectural Programming Primer.* Boston: Cahners Books International, 1977.

(Presented is a method which is structured to help professional designers work with clients.)

Preiser, Wolfgang F. E., Editor. *Facility Programming.* Stroudsburg, PA: Dowden, Hutchinson and Ross, 1979.

(This is a collection of papers on architectural programming by various authors.)

Preiser, Wolfgang F. E., Editor. *Programming the Built Environment.* New York: Van Nostrand Reinhold, 1985.

(This is a collection of papers on architectural programming by various authors.)

Sanoff, Henry. *Methods of Architectural programming.* Strousdburg, PA: Dowden, Hutchinson and Ross, 1977.

(The author reviews various techniques. Limited information is provided on each technique.)

White, Edward T., III, *Introduction to Architectural Programming.* Tucson, AZ: Architectural Media, 1972.

Industrial Facility Planning and Layout

These references are written primarily for industrial engineers involved in the planning of industrial buildings. Attention is given to deriving solutions to meet production and process requirements. Some are mathematical and computational in nature.

Francis, Richard L. and John A. White. *Facility Layout and Location: An Analytical Approach.* Englewood Cliffs, NJ: Prentice-Hall, 1974
(This is a mathematically based, optimization approach.)

Muther, Richard. *Systematic Layout Planning,* 2nd Edition. Boston: Cahners Books International, 1973.
(This is a long-standing reference for industrial facility planning and layout.)

Muther, Richard and John D. Wheeler. *Simplified Systematic Layout Planning.* Kansas City: Muther and Associates, 1962.
(This is a shortened version of the previous reference, intended to broaden application of the method to nonindustrial buildings.)

Muther, Richard and Lee Hales, *Systematic Planning of Industrial Facilities.* Kansas City, MO: Management and Industrial Research Publications, Vol. 1, 1979. Vol. 2, 1980.
(This is an expanded version of Systematic Layout Planning.)

Tompkins, James A. *Facility Design.* Raleigh, NC: North Carolina State University, 1977.
(This is a general treatise on industrial buildings.)

Tompkins, James A. and John White. *Facilities Planning.* New York: John Wiley & Sons, 1984.
(This book is aimed primarily at industrial, warehouse, and storage facilities. A section on materials handling is included.)

Building Design Standards

A wide variety of data has been compiled in these handbooks to help architects with designs. They may be helpful in URM for estimating space size for some types of space.

De Chiara, Joseph and John Hancock Callender. *Time-Saver Standards for Building Types.* New York: McGraw-Hill Book Co., 1973.

Ramsey, Charles G. and Harold R. Sleeper. *Architectural Graphic Standards.* New York: John Wiley & Sons, 1970.

Building Cost Estimates

These references are useful for estimating the cost of construction. Data are compiled from regional construction costs and continually updated. Much detail is provided for each component and phase of construction, such as foundation work, roofing, et cetera.

Building Systems Cost Guide. Kingston, MA: Robert Snow Means Co., Inc., annual update.

Dodge Manual for Building Construction Pricing and Scheduling. Princeton, NJ: McGraw-Hill Information Systems Co., annual update.

National Construction Estimator. Solana Beach, CA: Craftsman Book Co., annual update.

Facilities Management

These are some of the few publications for the rapidly emerging field of facilities management. Facilities managers are often involved in facilities planning.

Molnar, John. *Facilities Management Handbook.* New York: Van Nostrand Reinhold Co., 1983
(Emphasis is on the technical aspects of buildings and building systems, little on functional requirements.)

Facilities Design and Management. New York: Gralla Publications.
(This monthly magazine often contains useful articles on facility planning. Major emphasis is given to office buildings, but other building types are covered from time to time.)

Glossary

Activity An action by which a mission is accomplished; a subelement of a function. One of the Six Keys of Organizational Accomplishment. One of the PEAS of URM. An activity describes what a space is used for.

Budgeting The process of establishing how much money an organization will spend for different things during a period of time (typically on an annual basis). Planned expenditures for buildings, real property, or capital improvements are a major element of a budget.

Many organizations have standard or formal procedures for identifying and prioritizing potential building actions (lease, buy, or build) and deciding which ones will be funded. These procedures are also referred to as *budgeting* among facility planners.

Building subsystems The assemblies built into and procured with a building. Typical subsystems include structural, electrical, lighting, water, sewer, and HVAC (heating, ventilation, and air conditioning).

Code A formal set of rules about building design and construction. When adopted by government organizations, codes become laws. Examples of national codes are the National Fire Code and the National Electrical Code. Examples of state and local codes are building, zoning, plumbing, and ventilation codes.

Concept design The early phase of the building design process, sometimes considered the first 25 to 35 percent. It follows the very earliest design phase, called *schematic design,* and precedes the remainder of the process, called *final design.* In concept design attention is given to general organization of the building and its subsystems, site orientation, appearance, early calculations for subsystems and preliminary specifications.

Criteria Criteria are standards, rules, codes, or other forms of statements which are used by designers to satisfy building requirements, including user requirements. Some criteria are parts of laws and ordinances, others are found in designers' professional references and literature and are part of good design practice. Criteria are typically quantitative and often measureable.

Design-build The normal sequence of completing a building project: completing the design first and then proceeding with construction.

Equipment/supplies One of the Six Keys of Organizational Accomplishment; one of the PEAS of URM. In URM, equipment, furniture, supplies, and all the things which an organization brings into a building or space to assist in performing activities or to act on that is not acquired as part of the lease or construction project. Equipment and supplies do not refer to the subsystems and devices built into a building to support user activities.

Facilities management The orderly process of planning, evaluating, operating, maintaining, disposing of, acquiring, and replacing already built facilities. The application of management practices and strategies to buildings as a resources of an organization or company.

Fast track A sequence for getting a building designed and constructed faster than a design-build sequence by proceeding with construction as soon as designs of portions of the building and its subsystems are completed.

Feasibility Establishing whether a building project should move ahead or can be constructed, based on a study of potentially limiting factors like need, cost, technology, politics, et cetera.

Final design The latter portion of the design process (following concept design) in which details are attended to, final construction drawings and specifications are completed, and approvals and permits are obtained.

Flow chart A block diagram that illustrates the sequence of events, or of travel or movement of people, information, equipment, or assemblies.

Function A major action by which a mission is accomplished.

Massing The process of organizing large elements of building space into a single unit so that appearance, site orientation, and other factors are satisfied. A very early step in the building design process.

Mission The purpose for which an organizational unit exists; its primary goal or objective.

Objective In URM, this is a statement that explains why a facility project is undertaken and why URM is implemented.

Operations A term referring to the general process by which an organization conducts its business.

Organization A term used to refer to a collection of organizational units that make up a company, operating unit, or educational, religious, public, private, or other formal body.

Organizational structure The collection of organizational units arranged into a total organization.

Organizational unit An organizational unit is any formal or recognized element of an organization. Common terms associated with organizational units are department, division, branch, and section. One organizational unit may be composed of several other organizational units. However, each is a unit for URM.

PEAS An acronym for people or personnel (P), equipment and supplies (E), activities or operations, (A) and time or schedule (S), derived from the Six Keys of Organizational Accomplishment. The four kinds of data that form the basis for user requirements for buildings.

People/personnel One of the Six Keys of Organizational Accomplishment; one of the four PEAS of URM. In URM, people or personnel can be referred to by the numbers and types of skills or by names of individuals.

Planning The process of laying out a scheme, such as for solving a building problem.

Project Any kind of action taken to a) identify whether an existing building is inadequate or b) resolve deficiencies of existing buildings through lease, purchase, new construction, or modification of existing facilities.

Quantity of space The amount of space that a room or other space associated with a building consists of. Typical units of measure are square feet (area) or cubic feet (volume).

Quality of space The condition of a space created by characteristics or features, not including the amount or size of a space.

Reference time A moment in time for which URM data is compiled for an organization. It recognizes that an organization continually changes and is not static.

Requirement A statement that defines or explains what is needed. It does not normally describe a solution.

Requirements data Information about activities, personnel, equipment, and time that is used to justify or explain user requirements. Information which describes the contents, use, and occupants of a space. Part of URM data.

Examples:

People/Personnel:	*Equipment:*	*Activities:*
2 clerk-typists	*1 standard desk with chair*	*Word processing*
1 supervising engineer	*2 hydraulic presses (Model 19)*	*Tire mounting*
	8 parts bins (3′l x 4′w x3′h)	*Preparation of work requests*

Schedule/time One of the Six Keys of Organizational Accomplishment; one of the four PEAS of URM. The amount of time, utilization, or schedule of activities of users can be an important consideration for defining user requirements or creating solutions to building requirements.

Schematic design A very early sketch to illustrate general concepts for a design solution. The earliest phase in the design process.

Schematic drawings A diagrammatic representation, usually composed of symbolic elements, used to describe the structure and components of a building subsystem. A common form of documentation for concept design.

Six Keys of Organizational Accomplishment The six items which together facilitate an organization's achievement of its mission, goal, or purpose: people, activities, equipment, schedule/time, buildings/facilities, and funds.

Space Any portion of an entire building. There are interior and exterior spaces. Spaces are referred to as rooms, areas, work stations, hallways, and other names. One space may contain other spaces. A space may be bounded or may not be bounded by walls or partitions.

The term *space* is also used to refer to the quantity of space that a room or area consists of. Typical units of measure are square feet or square yards of floor area or cubic feet of volume.

Space management The application of management principles to an inventory of rooms and buildings to ensure that utilization is maximized and the correct amount and type of spaces are on inventory.

Space type Kind of space. Examples are: classrooms, medical examination rooms, private offices, and workshops.

Staff-facility specialist People on the staff of organizations or companies who are responsible for or involved with building and facility planning, analysis, assignments, and decisions; often located in an administrative or general services group. These specialists may also be called *facility managers* or *layout planners.* They may be professionals (architects, engineers, or other design and building technology professionals).

Standards A criterion or norm for measuring or allocating space quantity or quality.

Technical requirements These are requirements for a facility that deal with technical, legal, and other considerations. They ensure that a building can be constructed, that it will be in harmony with the surroundings (aesthetics, energy conservation, utility systems, roadways, et cetera), are reasonably safe for occupancy (fire protection, structurally sound, et cetera), are technically feasible, and will minimize operating and maintenance costs for the building itself.

URM The *user requirements method* for defining functional requirements for buildings and built facilities.

URM data The data compiled during the application of URM, consisting of user requirements and requirements data.

Users In URM, users are the occupants or users of an existing building or one being planned.

User requirement A requirement is a statement about what is needed. It is typically written in qualitative terms or in the form of a goal or objective to be achieved. User requirements are concerned with building characteristics that will allow users to perform their activities efficiently, safely, and with regard for occupant satisfaction. User requirements are called functional requirements by some.

Examples:

User requirements:

Thermal comfort conditions for occupants
Lighting for reading instruments on machines
Space to open access panel and service electronics rack

Appendix A
URM Forms

These forms are included in this book so that the reader can reproduce them when applying URM. Permission is granted to reproduce the blank forms in this appendix.

MISSION/FUNCTIONS

URM form A

organization

representative

personnel

male
female
total

mission

functions

subordinate organizational units

From: R. L. Brauer, *FACILITIES PLANNING–User Requirements Method,* American Management Association, 1986.

MISSION/FUNCTIONS

URM form **A**

Comments:

People Equipment Activities Schedule	URM form B

organizational unit	activities	comments
function	personnel	
	equipment	

From: R. L. Brauer, *FACILITIES PLANNING – User Requirements Method,* American Management Association, 1986.

Comments:

SPACES/ADJACENCIES

URM form C

organizational unit

PEAS	number	space name

adjacency codes

0–same space
1–must be adjacent
2–close, adjacency preferred
3–anywhere nearby
4–distance unimportant
5–must be distant

From: R. L. Brauer, *FACILITIES PLANNING–User Requirements Method,* American Management Association, 1986.

SPACES/ADJACENCIES

URM form C

Comments:

EQUIPMENT DATA

URM form **D**

organizational unit

equipment name	PEAS	conditions generated					conditions required					
		air contam	heat	noise	other	explanation	air cond	ventil	light	electr	other	explanation

From: R. L. Brauer, *FACILITIES PLANNING – User Requirements Method,* American Management Association, 1986.

EQUIPMENT DATA

URM form D

Comments:

REQUIREMENTS

URM form **E**

organizational unit		space name	
requirements		requirements	
Size	Method	Space Type	Class
need codes	N-absolutely necessary I-somewhat important L-like to have, if possible	purpose codes	1-activity or function 2-health or safety 3-morale 4-security 5-equipment 6-other

From: R. L. Brauer, *FACILITIES PLANNING – User Requirements Method,* American Management Association, 1986.

REQUIREMENTS

URM form E

Comments:

DIAGRAM

URM form **F**

type

- ◯ space relationship
- ◯ space layout
- ◯ flow chart
- ◯ organization chart
- ◯ organization relationships

title

From: R. L. Brauer, *FACILITIES PLANNING – User Requirements Method,* American Management Association, 1986.

DIAGRAM

URM form F

Comments:

SHARED SPACES

URM form G

organizational unit

space type	how often used	normal use time	usual and maximum number of people

comments

From: R. L. Brauer, *FACILITIES PLANNING – User Requirements Method,* American Management Association, 1986.

SHARED SPACES

URM form G

Comments:

FORECAST

URM form **H**

organizational unit	
when expected	
impacts on: type of space, amount of space, requirements	
nature of change	

From: R. L. Brauer, *FACILITIES PLANNING – User Requirements Method,* American Management Association, 1986.

FORECAST

URM form H

Comments:

From: R. L. Brauer, *FACILITIES PLANNING–User Requirements Method,* American Management Association, 1986.

URM form I

Comments:

SPACE & REQUIREMENTS	URM form J
organizational unit	space name

basis for size estimate (method, standards)	size (sq ft)	space number

requirements

layout sketch

From: R. L. Brauer, *FACILITIES PLANNING - User Requirements Method,* American Management Association, 1986.

SPACE & REQUIREMENTS

URM form **J**

Comments:

CONTENTS OF SPACE

URM form K

organizational unit	space name

activities

people & equipment

comments

From: R. L. Brauer, *FACILITIES PLANNING–User Requirements Method,* American Management Association, 1986.

CONTENTS OF SPACE

URM form K

Comments:

REVIEW COMMENTS

URM form **L**

reviewer		organizational unit	date
Reference: drawing/page/ space name or number	**comment number**	**comment**	

From: R. L. Brauer, *FACILITIES PLANNING – User Requirements Method,* American Management Association, 1986.

REVIEW COMMENTS

URM form L

Appendix B
Computers and URM

Appendix B

Computers and URM

Using Computers with Facility Planning

Computers and data management software can be a real asset in facilities planning. *However, computers will not help derive the data essential for good facilities planning.* They will help one sort through the data to find errors, to locate inequities in requirements, to easily find data needed for design solutions or evaluating alternatives, and to prepare standard or special reports quickly. They can simplify computations and reduce computational error.

Computers will not be very helpful for small building projects, but for larger projects a computer is essential. One cannot manage the mass of data by hand in a large project. The data cannot be manipulated or tabulated in all the ways needed to answer questions fast or easily enough.

The major benefits of using computers in URM are 1) saving clerical work, 2) saving time, and 3) making queries.

Forms are essential to collect requirements and supporting data in an organized manner. Initially, they have to be filled out by hand. Later, they may be typed for reports. In either case one has to make corrections, which involve erasures or redoing the page. When significant modifications must be made (a change in an organization, a reorganization, or a new operation), forms must be completely redone. With a computer, these changes can be made much more easily because one usually has an editing mode. Clerical work is greatly reduced when a computer is used.

The time required to process data is also reduced when a computer is used. Less time is needed to make changes and correct errors. The greatest time saving results from the ability to convert original data to a final document. In fact, the need to compile a written document can be eliminated because contents can be examined electronically. Some printed output will undoubtedly be needed, however.

Although clerical effort and time can be saved, the greatest benefit results from having a query capability. With a printed document, querying is often avoided because each search requires that one must page through each form and tabulate the data needed. The computer provides a powerful query capability that can be elicited by simple commands.

In one very large project in which requirements and supporting data was compiled in a computer, designers were able to answer at least 40 early design questions in less than two hours. Computer printouts provided the answers to such questions as "What are all the spaces among the half-million square feet in the building that require water?" or "Which spaces must have radio-frequency shielding?" A limited number of documents were made that contained all requirements. Each was about one foot thick. One can imagine trying to answer questions by having to hand-tally answers while paging through the book space by space.

Another organization compiled the administrative office space requirements for its entire staff of 2,500 people. A new office building project had to be defined within two months, but had to account for those organizations that would be placed in some renovated buildings. The difference between the total requirement and that solved by the renovation projects would be the requirement for the new building. Because data were compiled in a computer and query capability was available, the planning team easily found errors and inconsistencies in submitted requests and got them corrected quickly. Special requirements were easily analyzed to identify whether they would be accommodated by the renovated buildings. Capacities of renovated buildings could be matched to the needs of various organizations to establish a plan for assigning organizations to renovated buildings or the new one. Without a query capability, this could not have been accomplished within the time limits.

System Requirements

Today's microcomputers and software for them can meet the computer needs of URM, unless there are extremely large files of data. Then larger machines are needed. The kind of computer and features needed will vary with the size of a project and the kind of software available. For example, the administrative-office requirements project mentioned above was completed on a microcomputer with a 10-megabyte hard disk and a commercial data base management program (RBASE:5000 by Microrim, Inc.). The file was larger than 1.3 megabytes for 2,500 people. Additional files about retrofitted buildings and their capacity were created as well. For the half-million-square-foot project, a mainframe computer was used, primarily because it was all that the user organization had for which appropriate data-base management software (System 2000 for IBM mainframe computers) was available.

A key element is the software. Although filing programs could be used for URM, a fully relational data-base management system (DBMS) with multifile or relation capabilities is much better. Also essential are good report writing capabilities within the software and speed in processing large requirements files when queries are made. The commercial products in the marketplace are continually improving. Their capabilities in terms of speed, file size, and ease in making queries and creating reports are getting better. These factors must be considered when deciding whether an existing computer and software will work effectively for URM. To analyze specific program/machine combinations would be futile, since there are several hundred data-base management and filing systems available for micro-, mini-, and mainframe computers. Figure B-1 lists several factors to consider when using computers for URM.

Figure B-1. Checklist of considerations when using a computer for URM data.

TIME

How much time is allowed or available in the building project schedule to enter data, perform checks and analyses, and prepare reports?

> Although it may take some time to get computer files designed and data loaded, the time saved during querying and report preparation will most likely result in a net gain.

PROJECT DATA

How large will the project be? How many spaces or people will be included in data?

> A programmer will need to estimate file size and computer capacity.

What will the data be used for?

> Different uses may affect decisions about what computers or software will be used.

Will data be compiled and retained for several projects? How many projects? For how long?

> Answers to these questions will affect computer and software selection, based on their capacity.

Will highly formated reports be required? Can formats be flexible if software report-writing capabilities are limited? Must reports be presented on preprinted forms in fixed locations on the page? Will reports be used only by planners and designers for project analysis and design? Will reports be used for presentations and project promotion?

> Some software allows for development of reports directly on the monitor screen, while others require programmers to write a lot of code. Some software allows data to be entered on preprinted forms in the printer. Others do not. If a lot of different reports are needed, or if formats will change, then software with report-writer design aids will be essential.

COMPUTERS

Is a computer available? Will one be enough? Can it be dedicated to a building project for a period of time for data entry, analysis, or reporting? Must it be available for other uses at the same time? Will computer equipment have to be purchased or leased?

> The coordinator and assisting programmer will have to look into what is available and the time required to complete purchase procedures.

Will the computer be able to store all data for one or more projects without operators having to mainpulate disks or other storage devices?

> Selection of proper components (hard disk, floppy disk, tape drive, etc.) is usually based on the size of data and program files.

Will the computer be able to process large files in its active memory?

> One must know how much RAM (random access memory) the computer needs.

How long will a query take for a file when sorting is required? Without sorting?

> Sorting is one of the slowest DBMS operations. In part, the processing speed of the computer itself controls the sorting speed.

Will more than one input terminal or computer be needed to complete data entry? Will more than one be needed for review and analysis? Will the computer be needed at different locations?

> Portability or remote access and single- or multiple-user file systems are essential selection factors.

Will direct wire or modem data transfer capabilities be required?

> Availability and capacity (baud rate) of telephone or electronic modems must be considered.

What printers are available? Can they handle preprinted forms? How fast do they operate? Can they do graphics?

> The speed and capabilities of printers are important factors. They could limit what one can do with output and influence how the computer support files are designed.

SOFTWARE

Is software currently available? Will it have to be procured?

A computer can't operate without software or programs.

What skills are needed to use the software? Is it easy to use for data-base definition, data entry, query, analysis, and reporting?

The capabilities and features will determine how much the programmer will have to do, how many programmers will be needed, and when the support system can be ready.

Can it handle a building project? How many records can it handle? Will it have enough record length? Can it handle multiple relations or files? Can more than one file be open when a query or report is made?

Some DBMS programs can handle a limited amount of data records: others have virtually unlimited capacity.

How long does it take to process a query or report? How long does a sort take?

Some DBMS programs are very slow in query and sort modes.

Can input screens, report forms, and command menus be created? How long will it take to set up this data base with unique input and reporting capabilities defined?

The programmer must take up where the software doesn't already have necessary features.

Will data have to be transferred between programs on similar or different machines? Will special programs have to be written to handle data transfers?

Some software has limited file-transfer capabilities or unorthodox file formats that prohibit transfer to other machines and files.

DATA ENTRY STAFF

Will data be organized completely and edited before entry? Will entry clerks have to make judgments about validity of data or what to do with missing or questionable data?

> Clerks should not be forced to make judgments regarding data. Someone knowledgeable (most likely the coordinator) should handle editing of data. The coordinator may have to take the data back to a representative for clarification or correction.

Will data-entry clerks have to be trained for this task or on the use of the computer or software? Have they entered data for a previous building project?

> Experienced data-entry clerks are best. Training on URM data forms is essential.

PROGRAMMING AND OPERATING STAFF

Is someone available to define the data base? Do programmers know how to set up input screens or transfer files from tape or disk? Can they develop required reports? Will training be needed?

> A good, experienced programmer is essential for maintaining a URM schedule and to minimize redesigns.

Can planning staff and designers perform queries themselves? Will they need a trained operator to perform queries or generate reports? Will training be required for users?

> URM data is most effective in applications if it is easy for the coordinator, designers, or others to access and query data themselves.

APPLICATIONS

Will data be transferred to someone else, such as a design firm? Will they have the same hardware and software? Can data be transferred to their equipment and data base? Will the data base have to be set up on their system or can a copy of the set-up be transferred?

> Computer systems of all file users must be able to handle the software on which the files are built and be compatible with the original system and files. Otherwise, additional versions must be created. Eventual users and their systems will be important in selecting initial hardware and software.

Figure B-2. Structure of URM data elements.

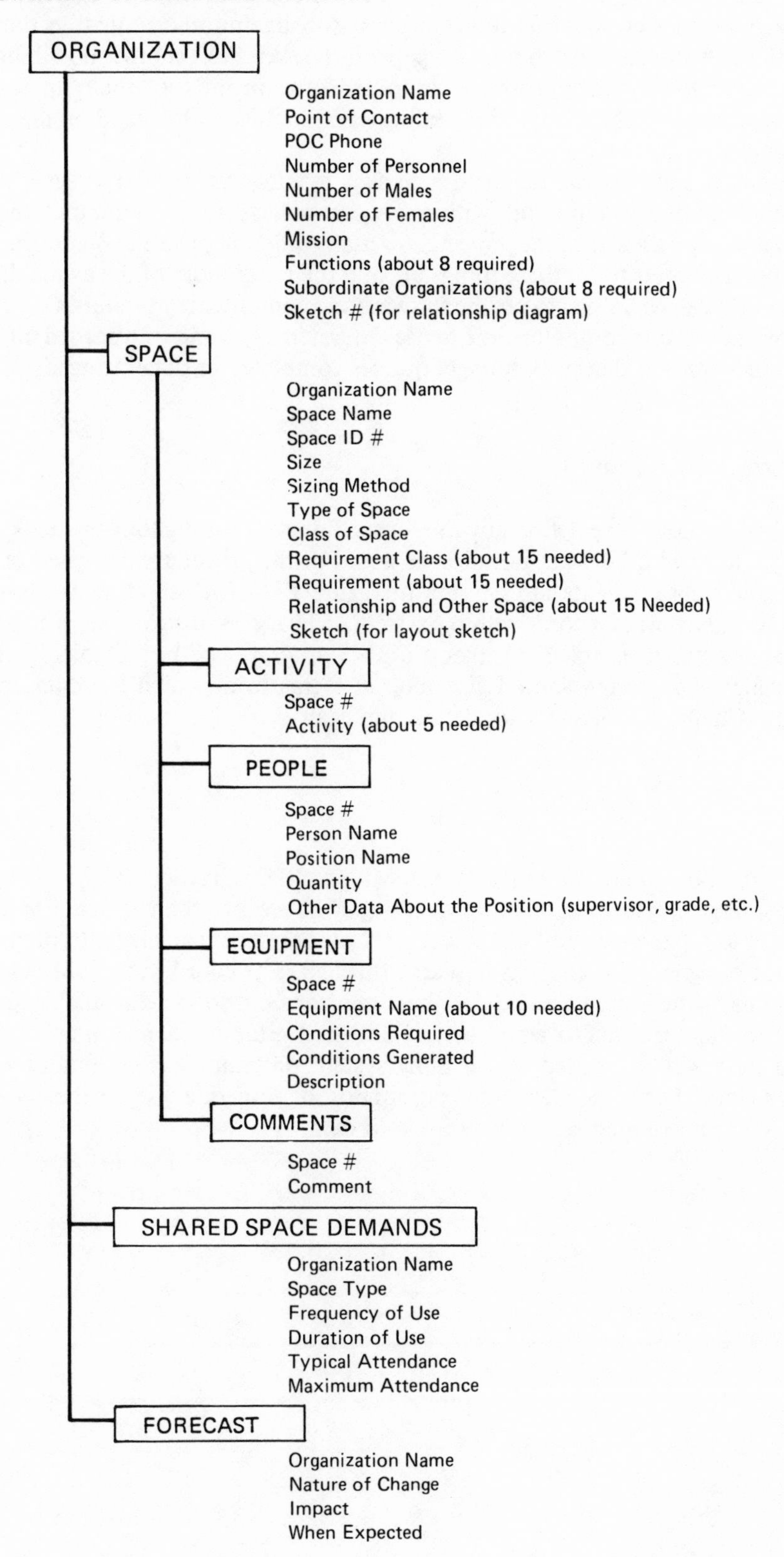

One thing that most filing and DBMS programs cannot handle is graphics. One way to deal with this is to give each sketch some identifying number and store that number with other data related to the sketch. Then, the only information to store on paper is the sketches. Another possibility is to convert sketches to electronic form with appropriate computer software and look them up or print a copy as needed from graphics files. The method for dealing with graphics will depend on the equipment and software available.

One can develop some special features, such as data-entry menus and screens, standard report formats, and standard query commands with many computer/software combinations. The value of doing this will depend on the size of the project, how many different projects will use the computer, the availability of programmers to set these things up, and the capabilities of the available software. For most projects people can be trained to use the software and computer commands. On one large project the organization trained one computer clerk to use the system. Anyone who needed information from the URM data files went to that person to get queries completed or data changed.

Data Structure and Elements

The data required by URM are defined by the forms discussed throughout the book. Data requirements will vary a little with URM application as well. For the procedure discussed in Chapter 2 the data structure and data elements are charted in Figure B-2. The structure is the same for each organizational unit. Layout of records within a system will vary because of system constraints. Figure B-2 will help programmers establish whether a DBMS program will be suitable. It will also help in designing data files. The design should also look at forms from which the data are derived (see Appendix A and Chapters 5, 6, and 7 for details and examples).

Summary

Computers and data-base management software will help the efficiency and effectiveness of URM. For medium and large projects, use of a computer should be seriously considered. For small projects it may be helpful, but a paper copy of URM data will probably be just as fast to prepare and not be unreasonable for designers and analysts to search through. The coordinator should locate someone who is very familiar with computers and data-base management software available and include that person on the coordinator's staff to get the system designed, established, and in use. A hierarchical or relational data base will be needed, not a filing system program. Actual loading of data can be completed by a clerk. The same clerk, the coordinator, or someone else on the coordinator's staff should be trained to make queries of and edit URM data.

Appendix C

Preparing Bubble Diagrams

Appendix C

Preparing Bubble Diagrams

Bubble Diagrams

Bubble diagrams are primarily a means for communicating relationships among spaces to designers. One of the first things a designer is concerned with in developing a solution to a facility problem is getting space organized. Initially, large blocks of space are manipulated. Then, elements within blocks and key relationships among particular spaces are resolved. In a new building, large blocks of space are organized to obtain an overall appearance and to achieve harmony with a site. This is called *massing*.

Bubble diagrams and the relationship matrices from which they are derived can also be used to evaluate a proposed design solution or layout.

Bubble diagrams are one step in the process of turning spaces, their size, and their spacial relationships into a two-dimensional layout. As illustrated in Figure C-1, adjacency data recorded on Form C is converted into bubble diagrams, then to block diagrams and, finally, to a final building plan.

The large units of space are represented in URM by the organizational relationship charts. They can also be composed by summing size across space type or classification (see Form E). These large units are combined with the relationship matrices (Form C) to establish a suitable solution. Each large block is organized using the data from the matrices that make up the block.

Purpose

The main purpose for preparing bubble diagrams in early planning is to help designers visualize what is represented by each relationship matrix. One could compile a large relationship matrix that includes every space in a building. However, it is very difficult to comprehend and keep track of all the relationships. Bubble diagrams simplify such a matrix and reduce it to essential relationships only.

Architecture and design disciplines related to it, such as interior design, are branches of fine and applied arts. Designers from these disciplines have a keen sense for the visual aspect of things and communicate using graphic media. Bubble diagrams are much more effective in communicating relationships to people with these backgrounds than are matrices of numbers.

Figure C-1. Relationship data recorded on Form C is translated into a building floor plan through intermediate bubble diagrams and block diagrams or block floor plans.

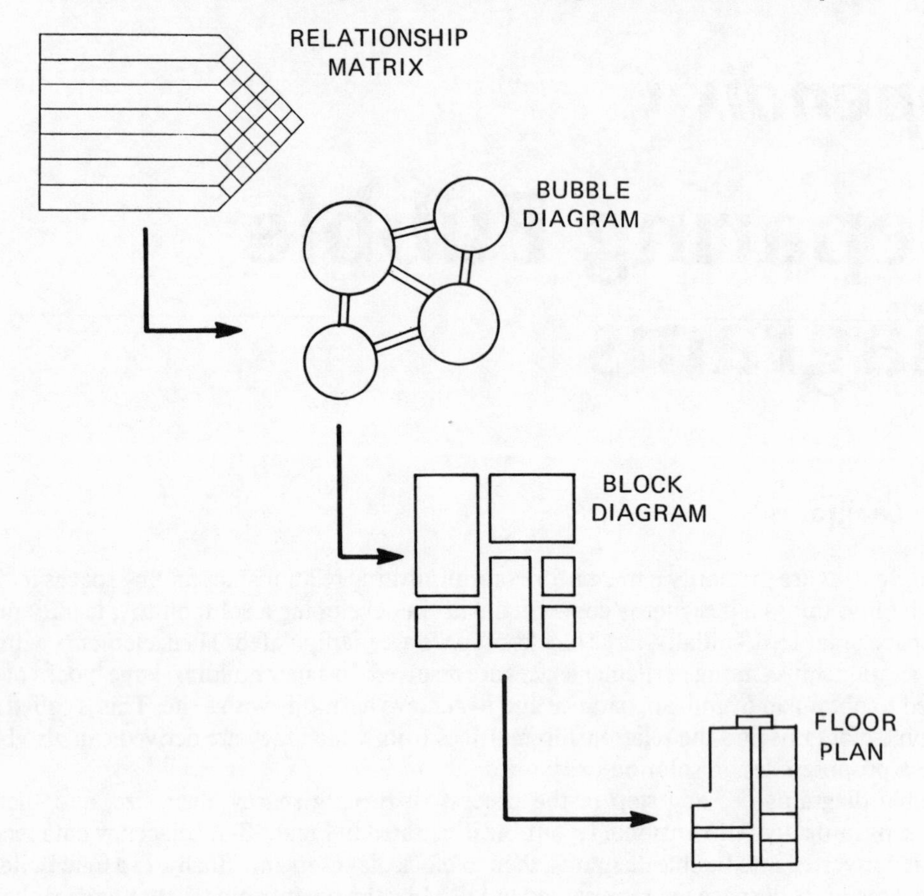

Limitations

While bubble diagrams can aid communication of key spacial and operational relationships, much of the data in a relationship matrix is lost when converted to a bubble diagram. However, this may not be too bad. The matrix data are merely an estimate of how things go together. Many considerations are reduced to a single number or letter that expresses the relationship between two spaces or organizations. Some factors that may be important are not presented precisely. In addition, when relationships are converted into a design, it is impossible to satisfy all relationships. A designer will make sure that all critical relationships (such as "1" or "5"; see Form C) are solved. Then, as many secondary ones ("2", "3," or "4") as possible will be resolved. The rest cannot be resolved precisely.

A relationship matrix and the bubble diagram that may result are merely guides. In many cases, it is a good idea to preserve the matrix with the bubble diagram that results, so that the fundamental relationship data is retained for evaluation of a proposed design. It is here that precision may be important.

Figure C-2. Circles in bubble diagrams can represent size of spaces through different computational schemes.

Original Data

ROOM	A	B
SIZE	4000 sq. ft.	300 sq. ft.

Scale Based on Circle Radius
(5000 sq. ft. = 1 in. radius)

RADIUS	$\frac{4000}{5000} = 0.8$ in.	$\frac{300}{5000} = 0.06$ in.
SCALED BUBBLE		

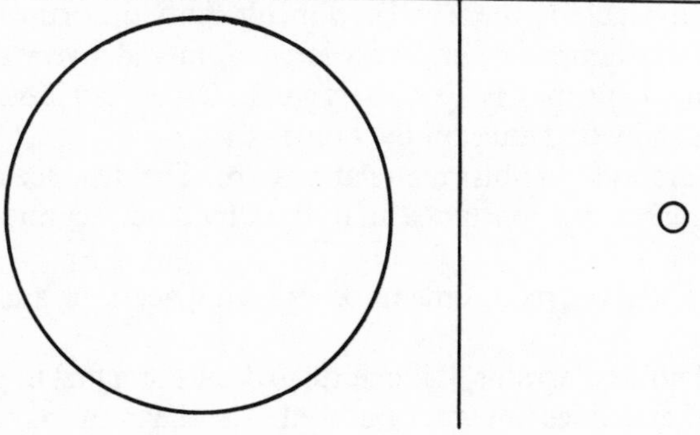

Scale Based on Area of Circle
(5000 sq. ft. = 3.14 sq. in. ; 1 in. radius)

AREA	$\frac{4000}{5000} \times 3.14 = 2.5$ sq. in.	$\frac{300}{5000} \times 3.14 = 0.188$ sq. in.
RADIUS	$\sqrt{\frac{2.5}{3.14}} = 0.89$ in.	$\sqrt{\frac{0.188}{3.14}} = 0.24$ in.
SCALED BUBBLE		

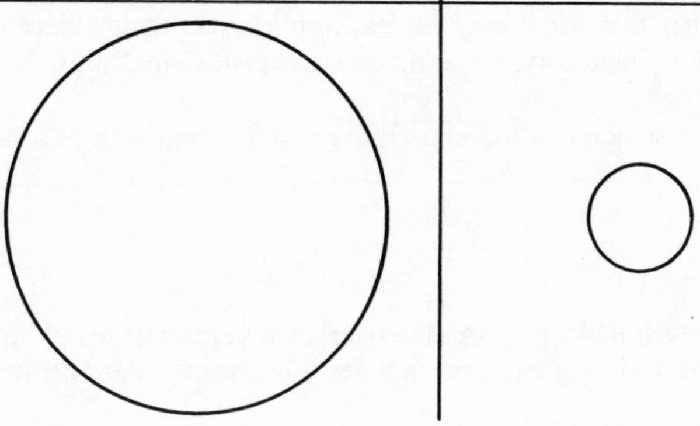

Bubble Diagram Methods

There are several methods for drawing bubble diagrams. Some are very simple, others more elaborate. Three bubble diagram techniques will be discussed below. They will be called *the distance method, the link method,* and *the annotated method.* A variety of artistic techniques can be used within each.

General Characteristics and Recommendations

In all bubble diagrams, each space (or possible organizational unit) is represented by a circle. The circles may all be the same size. Or circles can be drawn so that the size of each circle represents the relative size of the respective spaces. Circles can be grouped (see Figure C-2) to keep a diagram from getting too messy or cluttered.

When circle size is used to represent the size of a space, a formula will have to be developed to convert room size to circle size. The formula must be adjusted so that the circles will be a suitable size. If they are too large, a group of them will not fit on a drawing. If they are too small, they will not look good on the drawing and labeling them will be difficult. Sometimes the formula will have to be violated for artistic reasons. For example, when a very large room and a very small one are illustrated on the same diagram, the small one may be extremely small. Drawing it larger than called for by the scaling formula will make it show up better on the drawing.

Various means are used to illustrate relationships. The distance method uses distance between circles to represent adjacency importance. In the link method, a connecting link is used for adjacency information.

In the annotated method, requirements other than space, size, and adjacency are incorporated in the bubble diagram.

When drawing bubble diagrams, it is helpful to draw them first in pencil. Then they can be inked. Very often the first sketch does not work out well. The diagram may be unbalanced. Some bubbles may be squeezed together, while others have plenty of room. Connecting lines may be confusing or notes or labels too crowded.

Distance Method

In the distance method, each space (or organization) is represented by a bubble or circle. The most important or central space is located in the center of the diagram. Other bubbles are placed around it. The bubbles are also located relative to each other, using distance to illustrate adjacency.

In Figure C-3, bubbles are positioned so that they are touching if they must be adjacent (value-1 on Form C). They are placed farther apart if their adjacency is less important (value in Form C is 2 or greater). The farther they are apart, the less important is their adjacency (and the higher the value shown on Form C). Obviously, some precision is lost. Only "must be adjacent" is absolutely clear (circles are touching).

In URM the distance method is used to compile relationships among organizations.

Link Method

Another way to draw bubble diagrams is to start by organizing the right number and size of bubbles on a page. Then a link is drawn between bubbles to illustrate relationships.

Figure C-3. Distance between circles in bubble diagrams can represent strength of relationships.

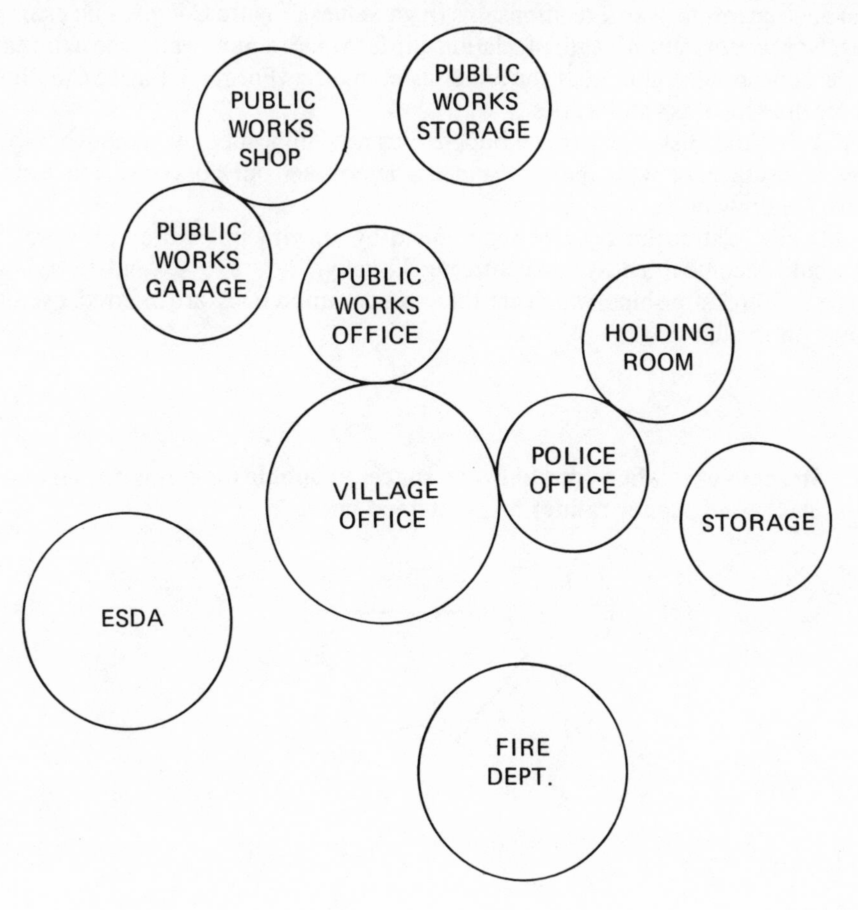

Figure C-4. Strength of relationships between spaces in bubble diagrams can be represented by lines of different thicknesses.

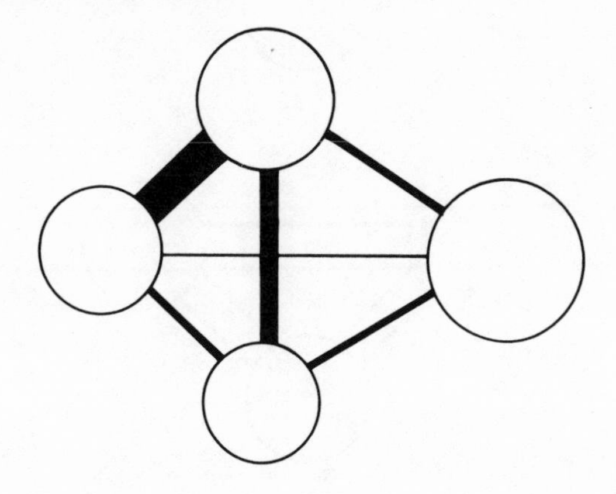

One way to represent strength of adjacency is to make the link width wide for strong relationships (low values) and narrow for weak relationships (high values). Figure C-4 gives an example.

Another way to represent strength of relationship is to place a number next to or on each link. The number is the appropriate value from the relationship matrix (Form C). Figure C-5 illustrates three techniques for drawing links and values.

In the link method, distance between bubbles has no significance, even though bubbles will end up with varying distances between them. Distance is determined only for convenience in fitting all the elements into the drawing.

Also, only key relationships can be represented by drawing links. If all links were drawn, the illustration would become too busy and confusing. Usually, "1," "2," "4," and "5" relationships are included. Then "3" relationships, which are the most common ones, are inferred, even though they are not shown on the drawing.

Figure C-5. Strength of relationships between spaces in bubble diagrams can be represented by placing adjacency ratings on connected lines.

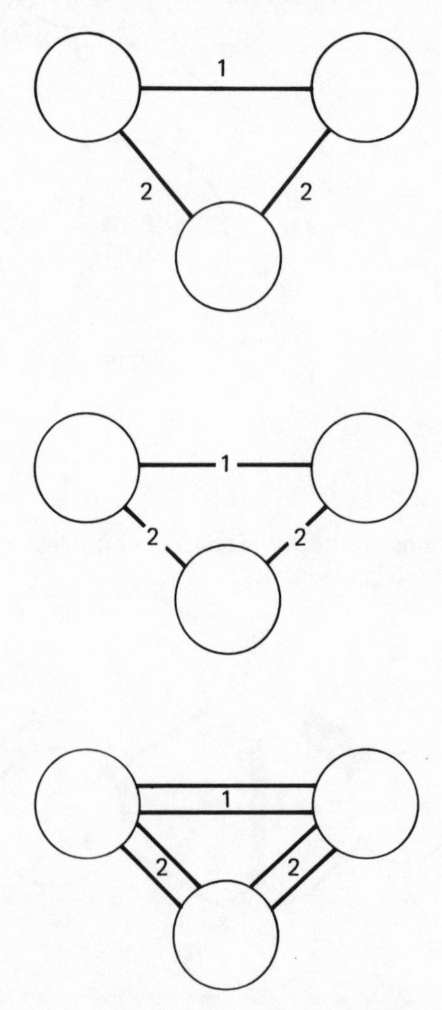

Annotated Method

In the annotated method, requirements other than space, size, and adjacency are included in bubble diagrams. Like space, size, and adjacency, additional requirements information is redundant to that logged elsewhere in written form in URM. However, converting requirements to a graphic, symbolic form may help the designers comprehend more easily or more completely what is needed. Therefore, the extra effort to communicate requirements in two modes may not be a waste of time.

Figure C-6. Symbols can be added to bubble diagrams to illustrate key requirements.

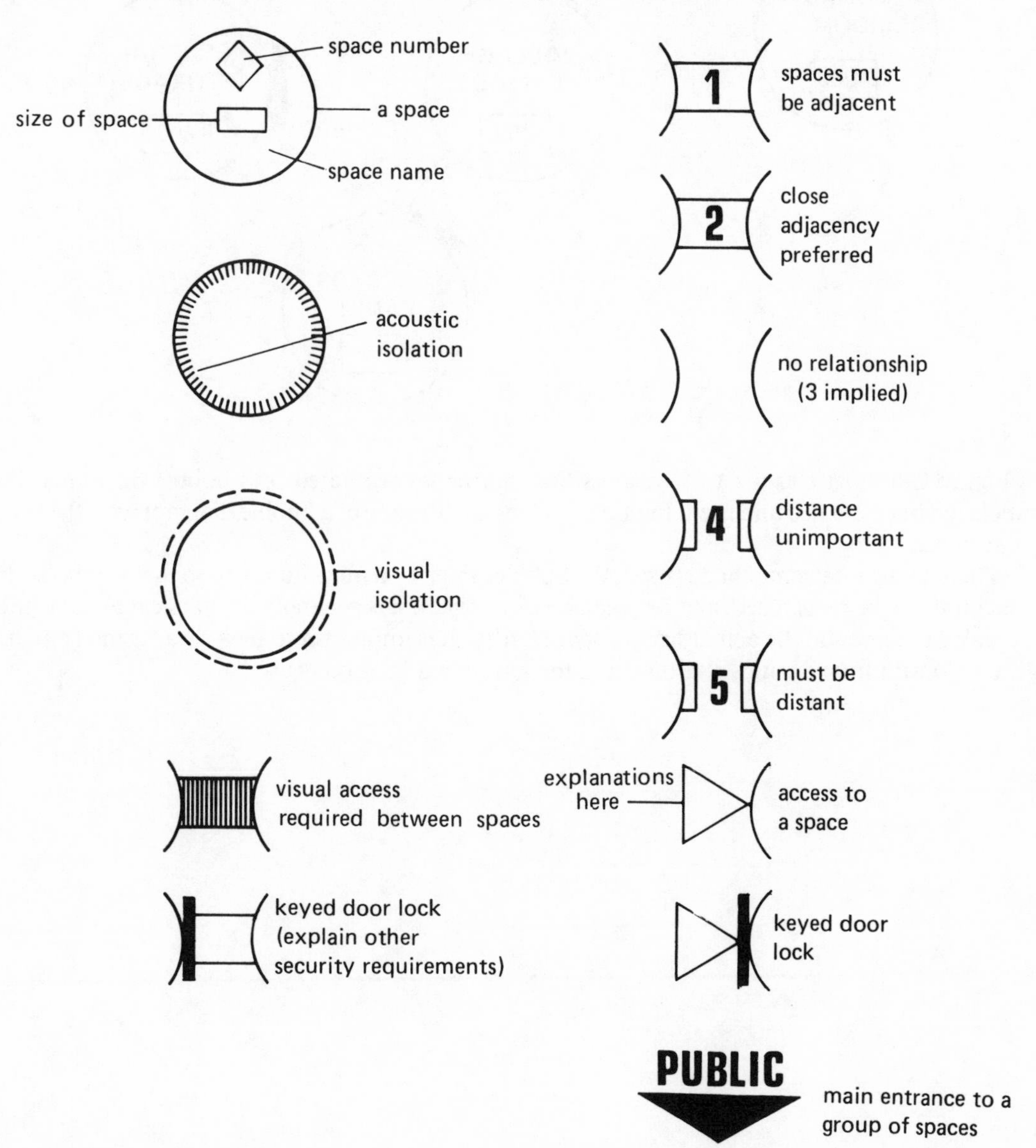

Figure C-7. Example of a bubble diagram with symbols included.

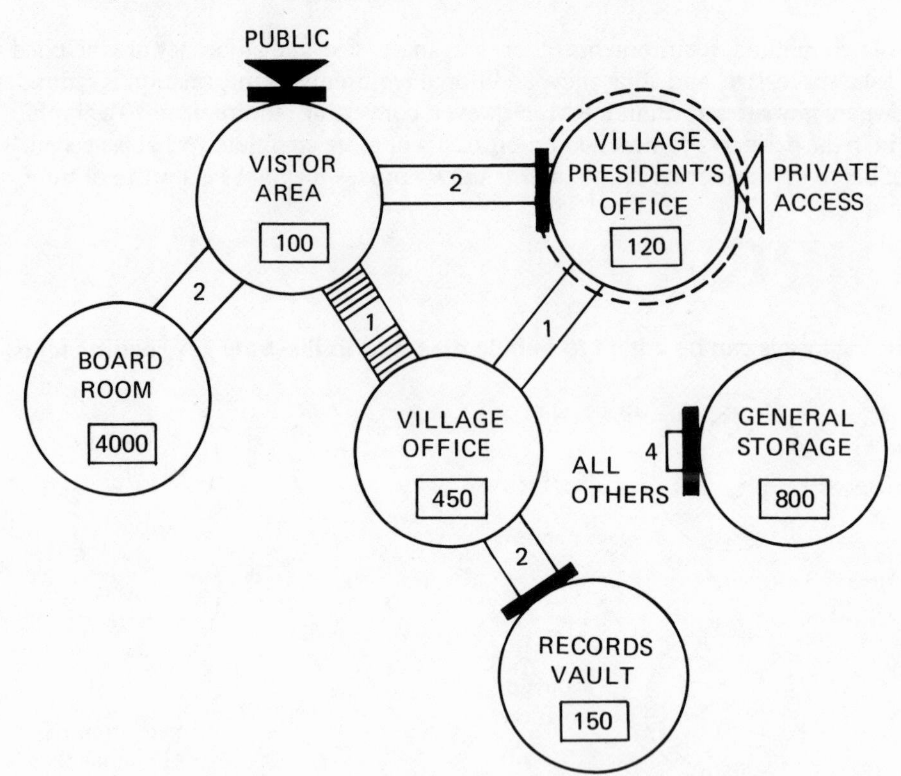

Figure C-6 illustrates a set of symbols that can be incorporated into bubble diagrams. Other symbols might be created and used. Figure C-7 gives an example of a bubble diagram with the symbol system in use.

When the annotated method is used, the bubbles must be a minimum size so that symbols can be placed on them, be recognized, and be readable. As a result, it is generally impractical to have bubble sizes scaled to represent the actual sizes of spaces. Also, it is important to include a legend (much like Figure C-6) with final documentation when the annotated method is used.

Appendix D

Project Data for Buildings

Appendix D
Project Data for Buildings

A project cannot move forward on the basis of user requirements alone. Other kinds of requirements and data are needed as well. Technical requirements must be formulated, funding requirements must be prepared, compliance with codes and standards must be checked, and other information must be developed and processed together with user requirements. Also, requirements which are unique to a project must be merged with general building information as the project moves through design and reaches a solution. Because many of the other requirements and data are formulated from or related to user requirements, it is important that user requirements be up to date and accurate.

The purpose of this appendix is to define project data and briefly discuss how it is related to user requirements.

Other Project Data

Besides user or functional requirements, the key data related to a project are technical requirements, cost data, criteria, codes, and standards.

Technical requirements are the collection of project requirements concerned with topics requiring specialized knowledge about buildings and their development. Technical requirements address such things as soils, drainage, roads, utility extensions, subsystem characteristics (like power, load factors, voltages, grounding, and lightning protection for electrical subsystems), and legal factors (like easements, pollution control, air-space rights, and building and occupancy permits).

Cost data has to do with project cost estimates, financing and tax strategies, bonds, payback period, return on investment, cost-benefit analysis, and similar information.

Criteria, codes, and standards are the national, state, local, organizational, or corporate laws, policies, and rules associated with building design.

Technical Requirements

Technical requirements are requirements for a building that ensure that it is safe for occupancy, has adequate supporting utilities and services, can be constructed, will have low operating and maintenance costs, and will be in harmony with its surroundings. Technical requirements are often derived from user requirements, but are not the direct concern of users. Technical requirements involve energy conservation, capacity of utility and waste systems, structural integrity, aesthetics of a building and its site, fire protection, materials selection, soil stability, stormwater drainage, durability and

Figure D-1. Items for which technical requirements and data are often developed.

General

Hazard and risk analysis
Installed equipment
Construction phasing

Site and Grounds

Location
Land use/zoning/air space/adjoining lands and uses
Highway, railroad, waterway, airport access, and right-of-way
Real estate transactions, convenants, restrictions or easements
Drainage and flood potential
Soils, boring data
Climate data
Construction restrictions and routes
Landscaping (lawn, ground cover, trees, shrubs)
Clearing, grading, and earthwork
Utility system access and capacities
Pavements (size and type)
Archeological and historical site
Security
Lighting

Structure

Seismic zone
Design loads (wind, snow, etc.)
Foundations
Floor and roof loads
Vibration isolation
Overall layout, massing, number of floors
Clear span
Security

Financial

Life cycle cost
Cost-benefit analysis
Return on investment
Cash flow
Payment schedules
Depreciation schedules
Tax credits

Energy, Electrical, and Communication Subsystems

Orientation and siting
Fuel source and type
Insulation
Fenestration
Power demands, peaks loads, outage acceptance
Heating and cooling loads
Control and energy management system
Electrical system
General and emergency lighting system
Telephone system
Intercom/paging/speaker system
TV system
Computer networks
Grounding and shielding

Mechanical and Environmental Subsystems

Heating system
Cooling system
Ventilation and exhaust system
Plumbing system
Special gases or liquids or compressed air
Water
Steam
Sewer and waste treatment
Solid waste
Security equipment and hardware
Vertical circulation (elevators, escalators)
Lift capabilities (cranes, hoists)

Fire Protection Subsystems

Occupancy type
Fire load and hazard analysis
Sensors and alarms
Extinguishing agents, extinguishers
Sprinkler and suppression systems
Site access
Water supply
Life safety
Explosives, flammables, and combustibles

capacity of roadways, and many other subjects. Technical requirements are the language of the building professionals (architects, landscape architects, engineers, and others) who have been trained to deal with these subjects. Technical requirements address many subjects for which laws have been enacted to provide for public safety and welfare.

Few users have training or experience sufficient for developing technical requirements for a building project or for dealing with them. Building professionals are usually employed to formulate technical requirements for a project. They may have to work with other professionals (such as attorneys) or with manufacturers, suppliers, and government agencies to define and resolve technical problems related to a project.

Because user requirements form the basis for many technical requirements (for example, the electrical power requirements for a building are derived, in part, by summing the power requirements for all electrical equipment of users), the coordinator and possibly some representatives may need to work closely with building professionals to help them translate user requirements into technical requirements. Representatives of users can help the building professionals extract data from user requirements and can assist in identifying impacts of technical requirements on user activities.

Figure D-1 is a list of items for which technical requirements and supporting data are typically compiled by professionals for projects.

Cost

Cost is a very important kind of data for building projects. It is the one Key of Organizational Accomplishment that allows the other five to exist or be effective in accomplishing an organizational goal or mission. Managers are paid to allocate funds to meet the greatest needs and accomplish the greatest effects. Few organizations have unlimited funds. Priorities are established and allocations made based on importance, return on investment, and other factors.

Feasibility studies may be needed to see what funding is required for a project, to determine whether the project or any of its parts will provide the benefits necessary to justify it, or establish its priority relative to other demands for funds. Costs and benefits for new construction or purchase may be compared to leasing. Financial planners, cost estimators, and other specialists are needed to develop funding requirements. To a great extent their work will build on user requirements data, such as total building size, special features, productivity, and organizational size.

Designers will use importance and need codes associated with user requirements to find the least-cost means for achieving an effective solution. User requirements data will help control project cost during design.

Criteria, Standards, and Codes

Criteria, standards, and codes are other forms of information important for building projects. These three terms are often used interchangeably and lack precise meanings that are acceptable to everyone.

In one sense, criteria refer to various forms of published data that are used by designers of a project to meet requirements. Designers work from many sources of published data that apply to buildings in general or to particular kinds of buildings or spaces. When used in this sense, criteria can include standards and codes. For some, a broad concept of criteria includes both user and technical requirements for a project.

In another sense, criteria are extensions of requirements. Using this meaning, criteria are the statements or references used by designers for satisfying requirements. In this sense, criteria are concerned with a design solution, not with what is needed. Criteria are stated in a form that can be measured and are usually quantitative. In contrast, requirements usually define qualities and are supported by requirements data (PEAS—see Chapter 2).

Standards, too, may be applicable to requirements or to design solutions, depending on the way in which they are expressed. For example, some organizations have developed space standards. Depending on the policy of the organization, these standards may be allowances (which make them requirements) or recommendations (which make them criteria). If space standards are allowances, a particular activity or person is authorized only so much space. This top-down approach is one way to define space requirements. URM is generally a bottom-up approach for defining space requirements. If space standards are recommendations, they can be used as guides in estimating how much space is needed during URM. Other kinds of standards may be used by designers as guides for formulating a design solution that will meet user requirements. Many reference books contain standards that are used by designers as guides for design solutions. Many organizations have established their own design standards.

Codes are design standards that have been formulated as recommended practice. Many codes deal with safety and health in building design and for the systems the buildings contain. There are fire codes, structural codes, plumbing codes, ventilation codes, codes for land use and building density, codes for fire protection, and codes for other subjects. When codes are adopted by federal, state, or local governments, they become law. Building professionals must have training and experience to apply codes effectively. In some cases, licensing is required to apply codes professionally.

Keeping User Requirements Current

Since other project data to a great extent depend on user requirements, one must keep user requirements current as long as information is being extracted from them.

Most organizations are continually changing. The Six Keys of Organizational Accomplishment (see Figure 2-3) provide a basis for recognizing change. The functions and activities of an organization may change. The number of activities may increase or decrease. The frequency of an activity may change. As a result, the number of people or the skills needed may change. The structure of an organization may be changed to improve performance. The equipment used may change because of technological advancement or the need to replace worn-out equipment. Time and scheduling, such as work hours or number of shifts, may change. Availability of funds may dictate change. Many kinds of changes for an organization can modify the user requirements formulated for a building during a single application of URM.

At least two ways for identifying the impact of change on user requirements were included in URM. One means was projecting changes. Another means for dealing with change in URM was selecting an appropriate reference time to capture effects on requirements for such things as an upcoming reorganization.

No one has 100 percent foresight. A sudden change in business climate or a technological breakthrough are typically not anticipated. In government, political restructuring of organizations that result from legislative action may be difficult to predict.

Normally, several months or even years elapse between the early stages of a building project and the time when occupancy can begin. During this time, many unforeseen changes are likely to occur. Changes must be monitored as the project progresses from early planning to lease or design and to final layout before occupancy. To insure that user requirements are as accurate as possible for a project, they should be periodically and systematically reviewed and updated. Although changes in design and construction may be needed as a result, the earlier the changes are introduced, the less costly they will be to implement.

Even after a building project (whether new construction, new lease, or a modification) is completed, changes will continue to occur in user requirements. As noted in Chapter 3, a periodic review of requirements is an important means for minimizing impacts of facility constraints on organizational accomplishment. By comparing updated user requirements to existing facilities, deficiencies can be detected and solutions initiated before things get so bad that deficiencies are recognized by everyone.

Index

About the Author

Dr. Roger L. Brauer was educated as a mechanical and industrial engineer. His special interest is in technology as it relates to people. Since receiving his doctorate in 1972, he has worked at the U.S. Army Corps of Engineers Construction Engineering Research Laboratory, and has focused on facilities planning and facilities management projects for various kinds of buildings. Much of his research work has been devoted to developing methods for getting facility users involved in building planning and design review, and in converting organizational and operational change into facility change. He has developed and tested methods for user formulation of facility requirements, user participation in review of designs, space management, and managing change in schools and training facilities. He has also developed methods for preparing design criteria that include typical user needs. He has published several articles on his work. Many of his methods have been incorporated in U.S. Army and government agency manuals.

Recently, Dr. Brauer has been developing microcomputer applications for facilities planning and management activities. He was selected the 1984 Researcher of the Year at the Construction Engineering Research Laboratory and he received the Albert Culbertson Award from the American Society of Safety Engineers.